From the Mind to the Feet

Assessing the Perception-to-Intent-to-Action Dynamic

Strategic Multilayer Assessment
Editorial Board

LAWRENCE A. KUZNAR
National Security Innovations (NSI), Inc.

ALLISON ASTORINO-COURTOIS
NSI, Inc.

SARAH CANNA
NSI, Inc.

February 2011

Muir S. Fairchild Research Information Center Cataloging Data

From the mind to the feet : assessing the perception-to-intent-to-action dynamic / [edited by] Lawrence A. Kuznar, Allison Astorino-Courtois, Sarah Canna.
    p. ; cm.
    Includes bibliographical references.
    ISBN: 978-1-58566-212-8
    1. Intention. 2. Decision making. 3. Situational awareness. 4. Military art and science – Decision making. I. Kuznar, Lawrence A. II. Astorino-Courtois, Allison. III. Canna, Sarah.

    355.02019--dc22

Air University Press
155 N. Twining Street
Maxwell AFB, AL 36112-6026
http://aupress.au.af.mil

Contents

PART 1
Operational Perspective:
Basic Issues in Gauging Intent

Foreword

From the Mind to the Feet: Assessing the Perception-to-Intent-to-Action Dynamic is an interagency, multidisciplinary collection of 12 essays addressing operational and academic perspectives on the elusive concept of an adversary's "intent"—its indicators and relation to behavior. It is primarily intended for the operational and policy community in the Department of Defense, the intelligence community, the Department of Homeland Security, and other US government agencies. The authors are from the intelligence community, the military services, US government agencies, federally funded research and development centers, academia, and the private sector.

The essays in this volume address the following set of critical questions:

- What do we mean by intent?
- How can intent be measured?
- What is the relationship of intent to behavior?

We developed the concept for this collection after completing a 2007 Strategic Multilayer Assessment (SMA) effort for Gen Robert Elder to operationalize the approach to deterrence described in the *Deterrence Operations Joint Operating Concept.** Part of that project involved a thorough analysis of how intent figures in the social fabric and how it influences an actor's decision calculus. Subsequent SMA projects elaborated on this work and highlighted the need for a more thorough consideration of what researchers and operators know about intent to act. SMA continues its work on deterrence and deterrence experimentation, and this volume is published as a supplement to these efforts.

---

*These essays were originally collected as a white paper—a product of the SMA effort. For those not familiar with SMA, it provides planning support to commands with complex operational imperatives requiring multiagency, multidisciplinary solutions that are not within core service/agency competencies. Solutions and participants are sought across the US government. SMA is accepted and synchronized by the Joint Staff and executed by the US Strategic Command's Global Innovation and Strategy Center (STRATCOM/GISC) and the Office of the Secretary of Defense, Director of Defense Research and Engineering, Rapid Reaction Technology Office (OSD/DDRE/RRTO).

These essays highlight three key observations:

- Despite near-universal agreement among academics, analysts, and operators that intent is essential, there exists no coherent body of research designed to address intent.
- Measuring intent requires multidisciplinary approaches involving psychology, neuroscience, decision theory, anthropology, and other social science disciplines, such as political science and sociology, that can establish the social context in which intentions form.
- There is a need for continued basic research to address the origin of intent and its relation to behavior and to develop complex models that capture how humans form intent and that can be used to analyze the masses of data required to gauge the intentions of individuals and groups.

While the short essays are written to stand alone and a selective reading would offer its own rewards, you are encouraged to read the whole report to gain the widest perspective on this critical issue.

I would like to take this opportunity to extend my thanks to the numerous contributors, the editorial board chaired by Lawrence A. Kuznar and Allison Astorino-Courtois, and Sarah Canna and April Hartman for compiling the manuscript.

Hriar Cabayan, PhD
Special Assistant
Office of the Secretary of Defense,
    Director, Defense
    Research & Engineering
    Rapid Fielding Directorate

About the Authors

Allison Astorino-Courtois is the chief decision sciences officer at National Security Innovations (NSI). She provides core support for a number of Department of Defense (DOD) Joint Staff and US Strategic Command (USSTRATCOM) Strategic Multilayer Assessment (SMA) projects, including a refocusing of DOD deterrence planning to the decision calculus of the actor(s) to be deterred, a decision and behavioral analysis of states and actors of particular interest to the defense community, and the design and production of an analysis tool for USSTRATCOM deterrence planners. Prior to joining NSI, Dr. Astorino-Courtois worked for Science Applications International Corporation (SAIC), serving as a USSTRATCOM liaison to US and international communities. She was also a tenured professor of international relations at Texas A&M University, where her research focused on the cognitive aspects of foreign policy decision making, and taught at Creighton University and the US Military Academy. She has received a number of academic grants and awards and has published articles in multiple peer-reviewed journals including *International Studies Quarterly*, *Journal of Conflict Resolution*, *Political Psychology*, *Journal of Politics and Conflict Management*, and *Peace Science*. She earned a PhD in international relations and an MA in research methodologies from New York University. She also earned a BA in political science from Boston College.

Elisa Jayne Bienenstock is the chief human sciences officer at NSI, where she is spearheading a number of DOD and intelligence community projects focused on developing and applying social network and decision-theoretic leadership models for anticipatory analysis and early indications and warning. Dr. Bienenstock received her PhD from the University of California, Los Angeles (UCLA), in mathematical sociology. She specializes in social psychology, with an emphasis on emergent properties of exchange. Her focus is applying formal models, such as social networks and game theory, to understand the relationships among elements of human interaction such as power, status, inequality, conflict, cooperation, coalition formation, reciprocity, and reputation. She is an adjunct professor at Georgetown

University where she teaches social network analysis for the Culture, Communication, and Technology Program. Prior to joining NSI, she was an associate at Booz Allen Hamilton. She has also held positions in the sociology departments at the University of California, Irvine; Stanford University; and the University of North Carolina at Chapel Hill.

CAPT John W. Bodnar, USN, retired, is a senior biological warfare analyst at SAIC in McLean, Virginia. His interest in modeling complex systems comes from previous experience conducting biological warfare analysis at the Defense Intelligence Agency, analyzing the revolution in military affairs as a Navy reservist for the Office of Naval Research and the US Naval War College, and researching and teaching bioinformatics at Northeastern University, the US Naval Academy, and Villa Julie College. Captain Bodnar has published in such journals as the *Naval War College Review*, *Armed Forces and Society*, and the *Marine Corps Gazette*.

Toby Bolsen is an assistant professor in the Department of Political Science at Georgia State University. His research focuses on the study of political behavior, preference formation, communications, and experimental methods. He has published articles in peer-reviewed journals such as the *American Journal of Political Science*, *Public Opinion Quarterly*, and the *Journal of Communication*. His work has also been supported by the National Science Foundation. He received his PhD in political science from Northwestern University.

Abigail J. C. Chapman provides innovative, research-driven solutions to the challenges in human, social, and cultural behavioral modeling for NSI. Prior to joining NSI, she worked for Defense Group Inc. (DGI), where she used findings from neuroscience to augment and refine social psychology findings on issues of concern to the military, intelligence, and national security communities. While at DGI, she designed an innovative multimethod approach using neurogenetics, social-cognitive psychology, and neuroscience to identify possible changes in neural anatomy, stress reactivity, and emotional and executive processing of stimuli in military personnel suffering from post-traumatic stress disorder and severe depression. Chapman has also worked at RAND's Intelligence Policy Center. She offers an international perspective gained through the course of extensive travel and study in Europe. Chapman earned her MA in applied social psychology and evaluation from Claremont Graduate University. Her BA is in international studies, with a focus on US foreign policy and European affairs, from American University.

Lt Gen Robert Elder, USAF, retired, joined the research faculty at George Mason University following his retirement from the Air Force as the 47th commander of Eighth Air Force. As the commander of Eighth Air Force and USSTRATCOM's Global Strike Component, he was responsible for nine wings and one direct reporting unit with 270 aircraft and 41,000 active duty, civilian, and reserve personnel. General Elder served as the first commander of Air Force Network Operations and led the development of the cyberspace mission for the Air Force. He also served as the air operations center commander and deputy air component commander for Operations Enduring Freedom and Iraqi Freedom. His executive experience includes senior leadership positions with the Joint Staff, Air Force staff, and NATO. His international experience includes interactions with senior government representatives in the Middle East, Asia, Europe, the Pacific, South America, and Africa. General Elder served as the commandant of Air War College and holds a doctorate in engineering from the University of Detroit.

Col Harry A. Foster, USAF, retired, is deputy director of the Center for Strategy and Technology at Air University, where he directs the Blue Horizons study, a Headquarters Air Force A8 and Air Force Research Lab–sponsored student study of the future of technology and strategy. During a 21-year Air Force career, Foster rose to the rank of colonel while gaining combat experience in the F-16, B-2, and B-1 during Operations Southern Watch, Allied Force, and Enduring Freedom. In staff positions, he led the US Central Command (USCENTCOM) branch of the Air Staff's Checkmate Division and later served as chief of strategy for USCENTCOM's air operations center. He holds master's degrees from the Harvard Kennedy School, Marine Command and Staff College, the School of Advanced Air and Space Studies, and Air War College.

Margaret G. Hermann is the Gerald B. and Daphna Cramer Professor of Global Affairs in the Department of Political Science and director of the Moynihan Institute of Global Affairs at the Maxwell School, Syracuse University. Her research focuses on leadership, decision making, crisis management, and comparative foreign policy. Dr. Hermann has been president of the International Society of Political Psychology and the International Studies Association as well as editor of the journals *Political Psychology* and *International Studies Review*. Among her publications are *Leaders, Groups, and Coalitions: Understanding the People and Processes in Foreign Policymaking* (with Joe Hagan), *Political Psychology as a Perspective on Politics, A Guide to Understanding Crisis Management through Case Studies* (with Bruce Dayton and Lina Svedin), "Using Content Analysis to Study Public Figures" (in Klotz and Prakash, *Qualitative Analysis in International Relations*), and "The Effects of Leaders and Leadership" (with Catherine Gerard in Dayton and Kriesberg, *Conflict Transformation and Peacebuilding*).

Kathleen L. Kiernan, a 29-year veteran of federal law enforcement, is the CEO of Kiernan Group Holdings, an international consulting firm which supports federal and civil clients. She is a senior fellow for the George Washington University Homeland Security Policy Institute and a faculty member at Johns Hopkins University and the Center for Homeland Defense and Security at the Naval Postgraduate School. She also serves as a special advisor to the director of the Combating Terrorism Task Force in the Department of Defense. Dr. Kiernan previously served as the assistant director for the Office of Strategic Intelligence and Information for the Bureau of Alcohol, Tobacco, Firearms, and Explosives (ATF), where she was responsible for the design and implementation of an intelligence-led organizational strategy for explosives, firearms, and illegal tobacco diversion. She also served as the ATF representative to the Counterterrorism Center at the Central Intelligence Agency (CIA). She has a doctorate in education from Northern Illinois University and a master of science degree in strategic intelligence from the Joint Military Intelligence College in Washington, DC. She also holds a master of arts degree in international transactions from George Mason University.

Lawrence A. Kuznar, the chief cultural sciences officer at NSI, is currently supporting various SMA projects for the Department of Defense, including the development of a sociocultural system typology to enhance military and intelligence analysis and planning efforts. He supports NSI efforts through development of computational models and supports human terrain system capabilities through development of social typologies and a review of ethics issues regarding the conduct of social science in a military context. Prior to joining NSI, Dr. Kuznar was a tenured professor of anthropology at Indiana University–Purdue University, Fort Wayne. His research focused on decision theory, theories of conflict and terrorism, and computational modeling. He conducted extensive research among the Aymara of southern Peru and with the Navajo in the American Southwest. He has published and edited several books and numerous peer-reviewed articles in journals such as *American Anthropologist, Current Anthropology, Social Science Computer Review, Political Studies, Field Methods*, and *Journal of Anthropological Research*. Dr. Kuznar earned his PhD and MA in anthropology and an MS in mathematical methods in the social sciences from Northwestern University. He holds a BA in anthropology from Penn State.

Daniel J. Mabrey is an assistant professor in the Henry C. Lee College of Criminal Justice and Forensic Science at the University of New Haven in Connecticut. In addition to his academic position, Dr. Mabrey is the executive director of the Institute for the Study of Violent Groups, which is a federally funded open-source research and analysis institute focusing on terrorism, extremism, and transnational crime groups and activities. He has authored several peer-reviewed journal articles and books including *Homeland Security: An Introduction*, published by Lexis-Nexis. He earned his PhD in criminal justice from Sam Houston State University.

Sabrina J. Pagano has experience in academic and private-sector settings as a researcher, instructor, and acting director of a growing behavioral sciences program. She has provided topical, methodological, and statistical expertise to support a wide variety of projects in areas ranging from message propagation and trust to the measurement of progress in conflict environments. In her transition from academia to industry, Dr. Pagano served a one-year appointment as a faculty fellow researcher and lecturer at UCLA in the Department of Psychology. In that position, she led the social psychology arm of a large interdisciplinary project investigating deficits in sociomoral emotions in frontotemporal dementia, a neurodegenerative disease. Dr. Pagano earned her PhD and MA in social psychology, with a minor in measurement and psychometrics, from UCLA. Her research focused primarily on the relationship between people's experience of distinct emotions and their support for distinct political or prosocial actions. She also holds a dual BA in psychology and political science from the University of North Carolina at Chapel Hill.

Thomas Rieger began working with international polling data and predictive models in 1986, after receiving an MS in industrial administration from Carnegie Mellon University. Since 2001 he has conducted extensive research regarding the reasons why complex systems destabilize, within both organizations and societies. From this work, he developed the Gallup Leading Assessment of State Stability (GLASS) model of state stability and the Political Radicals (POLRAD) model of political radicalization. He has applied these and other models to a number of defense-related efforts in several theaters and has consulted on developing frameworks for reconstruction and stability. Rieger currently leads global consulting efforts regarding organizational and societal barriers for Gallup Consulting and has spoken and written extensively on the topic.

Gary Schaub, Jr. is an assistant professor in the Leadership and Strategy Department of Air War College, Maxwell Air Force Base, AL. He previously taught at the Air Force's School of Advanced Air and Space Studies, the University of Pittsburgh, and Chatham College. Dr. Schaub has published articles in the journals *International Studies Review*, *Political Psychology*, *Parameters*, *Strategic Studies Quarterly*, and *Refuge*, book chapters in volumes published by Oxford University Press and St. Martin's Press, as well as op-eds in the *New York Times*, *Washington Times*, *Pittsburgh Post-Gazette*, and *Air Force Times*. He has written on military coercion, nuclear weapons issues (arms control, doctrine, and proliferation), European security, and research methodology. His current research addresses theories of coercion, decision making, and civil-military relations.

Keren Yarhi-Milo is an assistant professor in the Department of Politics and the Woodrow Wilson School of Public and International Affairs at Princeton University. Before joining Princeton in 2009, she was a postdoctoral fellow at Harvard University's Belfer Center for Science and International Affairs. For the 2007–2008 academic year, Yarhi-Milo was a fellow at the Olin Institute for Strategic Studies at Harvard University. She has worked at the Mission of Israel to the United Nations and served in the Israel Defense Forces, Intelligence Branch. She has received awards for the study of political science from the Smith Richardson Foundation, the Arthur Ross Foundation, and the Christopher Browne Center for International Politics. Dr. Yarhi-Milo holds a PhD and a master's degree from the University of Pennsylvania and a BA in political science from Columbia University.

Preface

Lt Gen Robert Elder, USAF, Retired

If you know the enemy and know yourself, you need not fear the result of a hundred battles. If you know yourself but not the enemy, for every victory gained you will also suffer a defeat. If you know neither the enemy nor yourself, you will succumb in every battle.

—Sun Tzu
The Art of War

The intent to act connects each actor's ideology and world-view to his or her behaviors. This collection of papers addresses the necessity of understanding the dynamic process through which perceptions are formed, how perceptions evolve to intent, and how intent becomes actionable. This dynamic is important, not only because it is impossible to influence an actor that you do not understand, but also because we must understand the causal relationships between our own perceptions, intentions, and actions.

Threat is not only a function of capabilities, but also intent and opportunity. Using technology and intelligence expertise developed over many years, we do a relatively good job estimating adversary capabilities; we are not nearly as good at estimating the intent of state or nonstate actors. Our capacity to influence others through kinetic and nonkinetic means rests on a multi-disciplinary understanding of the perception-to-action dynamic.

Intelligence analysts integrate data and information to develop accurate, timely, complete, and actionable intelligence estimates. Technical collection products include facts regarding the terrain, economies, weapons systems, military capabilities, and operating concepts of state and nonstate actors. Although human intelligence provides reporting on leadership profiles and intent, it is not always available in denied areas. Additionally, without verification by source data, it is merely human reporting rather than integrated intelligence. Our ability to portray leadership intent and predict leadership behavior is impeded by these intelligence gaps.

What connects an actor's ideology to his intent, and what connects his intent to his actions? Neurologists, sociologists, cultural anthropologists, and linguists attest to the complex relationship between the mind and the body, between culture and perception, between perception and intent, and between intent to act and action. This collection is an exploration of these dynamic steps and provides a meaningful theoretical and applied discussion of these concepts. The authors, both academicians and military subject matter experts, provide invaluable insight by adding their accumulated knowledge and expertise to the project. For this we are extremely grateful.

Executive Summary

Allison Astorino-Courtois and Lawrence Kuznar, NSI, Inc.

The purpose of *From the Mind to the Feet* is to open a much deeper dialogue about gauging intent than currently exists either in operational or academic arenas. It is intended to serve military and civilian defense leaders, deterrence and policy planners, and practitioners with a review of the basic concepts and state-of-the-art understanding of *intent*. It is organized into two sections: operational and academic perspectives on intent.

Operational Perspective: Basic Issues in Gauging Intent

The first section of this volume consists of four operational perspectives on intent. Kathleen Kiernan and Daniel J. Mabrey offer an enlightening description of law enforcement "street-craft" and explain how police methods of assessing an individual's intent to offend can inform counterterrorist operations.

Harry Foster discusses the ways military planners typically assess the intent of state or nonstate leaders and then offers what he calls an "effects-based thinking" framework for measuring intent. Foster argues that this effects-based thinking combined with a betting methodology offers the best analytic framework for melding the analyst's intuition (i.e., the art) with analytics to gauge intent.

Gary Schaub, Jr., widens the aperture to focus on how to gauge the intent of groups and large collectives such as nation-state and nonstate actors in the context of strategic deterrence. Schaub argues that a process to infer adversary intent on a continuous basis is needed so that a serviceable product is available to assist both in routine and crisis planning. He posits analysis of competing hypotheses as one method of achieving this goal that encourages debate and critical thinking about the adversary to prevent the typical intelligence errors of mirror imaging or groupthink.

Finally, John Bodnar outlines a new way of thinking about adversaries—and our relationship to them—in the twenty-first century.

Academic Perspective: Theory and Research in Gauging Intent

The second section of this volume consists of perspectives on intent that represent research and theoretical work in seven academic disciplines: anthropology, social psychology, international politics, social cognitive neuroscience, survey science, communications, and decision science.

In the first piece in this section, Lawrence Kuznar provides a comprehensive review of basic motivating factors recognized by anthropologists that help explain intent, including structuralism, interpretivism/symbolic anthropology, postmodernism, culture and personality, human behavioral ecology, and discourse analysis. It provides insight on intent as derived from socially shared systems of meaning and ideology.

Margaret Hermann concludes that learning how policy makers view what is happening to them is critical to understanding how governments are likely to act. With our unprecedented access to policy makers' perspectives through 24-hour news cycles, new ways to measure and interpret these perspectives are being developed. Hermann's assertions are tested by a case study in the following chapter.

Based on international relations research, Keren Yarhi-Milo provides analytic support for what we have always suspected: intelligence analysts and decision makers do not view the world or think about it in the same ways. Yarhi-Milo uses three test cases: British assessments of Nazi intentions prior to World War II, the Carter administration's assessments of Soviet intentions, and the Reagan administration's assessment of Soviet intentions in the final years of the Cold War.

Neuroscientists are actively unlocking the inner workings of the brain and revealing how goal-oriented, deliberative behavior interacts with emotive impulses to generate intentional behavior. In companion pieces, social neuropsychologists Sabrina Pagano and Abigail Chapman shed light on the latest in brain studies and the possible future applications of insights gained from neuropsychology for estimating and capturing an individual's intent to act in a certain way.

Tom Rieger offers empirical evidence that large-scale instability-based violence and ideologically based violence are driven by different factors with different purposes. Iraq is used as an example to explain extremist violence and to illustrate how it is possible to win a war and lose the peace.

Representing social communication and messaging, Toby Bolsen discusses information-framing research and the gaps in the literature regarding the links between manipulations of perceptions and attitudes (framing) and short- and long-term behavior.

Finally, theories of intent abound, but the most difficult task is the measurement and assessment of this intangible. Elisa Bienenstock and Allison Astorino-Courtois address some contemporary efforts in actually measuring intent and propose a backward induction method for honing in on intent based on behavioral "probes" identified through decision analyses.

Introduction

What is intent? Most of us can answer this question easily: it is a determination to act in a certain way for a certain purpose. It is a mental construct. But how do we understand or measure someone's intent to act? This is a trickier question. So tricky in fact that even in complex analyses, it is often assumed away—as in analyses based on the concept of a profit-motivated "economic man" or alternative models of nonrational decision making[1]—inferred from circumstances, taken as given from self-reports (conclusions drawn from preelection polls), or attributed based on our own perceptions (e.g., Khrushchev put missiles in Cuba to be poised to attack the United States).

Many social scientific and military disciplines—psychology, political science, international affairs, economics, linguistics, anthropology, intelligence, and operational studies—address intent. Nevertheless, as noted, intent as a concept remains difficult to define in a clear and unambiguous manner. It is even more challenging to measure. This volume contains contributions from authors representing multiple academic disciplines, analytic approaches, and security-related experience and training. It is intended to serve military and civilian defense, deterrence, and policy planners and practitioners with a review of the basic concepts and state-of-the-art understanding of intent. This volume may also inform the broader academic community by assembling a diverse set of research disciplines as well as the experience of practitioners.

Precisely because intent is the mental tie that links perception and behavior, it is a key component of most national security threat equations. The central thesis of this volume is simply this: understanding, estimating, and even measuring intent are critical and should not be given short shrift, especially in the areas of national security and defense. Failing to account for intent in a meaningful way amounts to ignoring half of what constitutes a threat in the simple but common additive model where *threat = intent + capability*. The issue is even more acute in the case of multiplicative threat models (i.e., *threat = intent * capability * opportunity*), where the intent factor can entirely determine whether or not a threat exists.

1

The papers in *From the Mind to the Feet* provide fresh and informative perspectives on the conceptualization and measurement of intent. However, they are just as valuable for what they lack. Developing robust theories and valid measures of intent requires a deep research tradition along with well-developed applications. This has not yet occurred. There currently is no community of "intent" scholars, for example, the way there is a well-defined, multidiscipline community of terrorism scholars. There is not an accepted methodology for measuring intent. A telling indicator of this lack of community and deep body of research is the fact that many researchers we contacted had difficulty seeing the connection of their research to the notion of intent. Intent is not yet a field of study nor even a buzzword in the academic community.

This volume is organized into two sections: operational and academic perspectives on intent. In accordance with the notion that the best way to learn is by doing, we begin with operators and practitioners describing contemporary perspectives on and lessons learned about the role of intent and the challenges in measuring it in military, intelligence, and law enforcement settings. The second section presents perspectives from various academic disciplines often used to inform operational efforts to deal with intent. The approaches and perspectives included in this volume represent approaches derived from:

- anthropology,
- social psychology,
- decision science,
- international politics,
- social cognitive neuroscience,
- survey science, and
- communications.

The purpose of this volume is to open a much deeper dialogue about gauging intent. In addition to delineating some of the existing research and ideas, the contributors have provided powerful indicators of the scope of the research yet to be done. Thus we hope that, as well as informing discussion and practice, the essays can serve as a guide for further government and academic research. We hope you will find each piece thought provoking and helpful.

Note

1. Economic man is the concept of humans as rational and broadly self-interested actors who have the ability to make judgments towards their subjectively defined ends as first described by John Stuart Mill. Nonrational models include bounded-rationality models and prospect-theory models, in which decision makers utilize simple heuristics instead of rational cost-benefit analyses.

PART 1

Operational Perspective
Basic Issues in Gauging Intent

Chapter 1

From Shoe Leather to Satellites

Shifting the Conceptual Lens

Kathleen L. Kiernan, EdD, Faculty, Johns Hopkins University and Center for Homeland Defense and Security, Naval Postgraduate School

Daniel J. Mabrey, PhD, Assistant Professor, University of New Haven, Henry C. Lee College of Criminal Justice and Forensic Science

Abstract: There is a corpus of experience in the law enforcement profession of dealing with criminals and criminal behavior, which, when understood in the context of support to and facilitation of terrorist activity, can help the defense community understand adversarial intent. Law enforcement streetcraft reveals that criminals have common motivators in which patterns of activity emerge that are known and measurable. These patterns contain embedded signatures which, when analytically unraveled, reveal the tactics, techniques, and procedures (TTP) of the individuals or groups responsible for the illegal activity. Rational-choice-based approaches in law enforcement to recognizing a criminal's intent are routinely used to help understand and anticipate criminality. Modus operandi analysis also assists in determining intent in that it forms the basis of many pattern/signature matching approaches used to identify suspects in an investigation. While previous patterns of behavior have not been conclusively proven to ensure accurate prediction of future behavior or intent, they do provide indicators and can contribute to analysis and moving the decision cycle "to the left," before intent becomes action.

> *Knowing the place and the time of the coming battle, we may*
> *concentrate from the greatest distances in order to fight.*

—Sun Tzu
The Art of War

Understanding the intention of a distant adversary is a requirement that has perplexed the intelligence, military, and law enforcement communities throughout history regardless of how proximate or well known the adversaries and battleground are. Understanding intent becomes even more complex with the advent of new adversarial tactics and techniques, such as cyber attacks, where the battlespace is largely unknown and ungoverned, where there are no clear rules of engagement, and where the adversary may disguise not only its intent but also its very existence.[1]

The purpose of this paper is to suggest that there is a corpus of experience in the law enforcement profession of dealing with criminals and criminal behavior which, when understood in the context of support to and facilitation of terrorist activity, can contribute to the knowledge of the military. An underlying premise is that, while every criminal is not a terrorist, every terrorist is in fact a criminal who must by necessity employ tactics which parallel those in the criminal world to include weapons acquisition, financing, false documents, sanctuary, and support. The support may manifest as an underground economy which transcends culture, geography, and routes of passage for the smuggling or concealment of people and commodities. The actual routes are static, perhaps enhanced through the introduction of technology such as sensors or lighting in tunnels or semi-submersible submarines vice go-fast boats. The actors and the commodities may change as well, but the intent and drive towards profit for the criminal world and the use of profit to fuel other terrorist-related activities do not. The examination will include the cultural and contextual expertise which enables skilled law enforcement practitioners to discern what is hidden in plain sight and invisible to the untrained eye. This experience or *streetcraft* is continually refined by practitioners in their persistent interaction with and observation of human behavior for which a common motivator is profit and in which patterns of activity emerge that are known and measurable.[2]

These patterns contain embedded signatures which, when analytically unraveled, reveal the tactics, techniques, and procedures (TTP) of the individuals or groups responsible for the illegal activity in a temporal as well as visual manner. In the early days of organized policing, rudimentary efforts included the accumulation of massive quantities of data in paper form, and often individuals hoarded knowledge and confined it to their memories as a means of gaining power or status. Unfortunately, important data was lost upon retirements or changes in assignment. While information sharing presents another complex problem, advances in technology have made it possible to archive historical, as well as current, data and to utilize a variety of sorting, matching, and analytical tool suites to interrogate the data. The first iterations of this were basic computer-aided dispatch systems and records management systems. Today policing is approaching the state of the art with integrated technology suites combining administrative, investigative, intelligence, threat/risk assessment, and resource management capabilities federated within and between organizations and jurisdictions. Leading technologies in this area are platforms and systems built by vendors such as COPLINK, Memex, and Tiburon.

Investigative techniques native to law enforcement are now being blended with analytical techniques that traditionally were the domain of the intelligence world to improve the efficiency of law enforcement operations. Geospatial analysis of crime and criminal behavior has been a mainstream application in law enforcement for more than 15 years and has driven military and intelligence applications of this technology. One such application, geographic profiling of an individual actor's behavior,[3] has demonstrated real-world success in understanding pattern-based crimes and apprehending serial offenders.[4] Link analysis, originally an application and method of intelligence analysis, is now routinely used by police departments throughout the United States to organize information collected in criminal investigations.

While the sensational crimes of serial murder, mass murder, and sexual predatory behaviors are often discussed in the psychological context of compulsion, antisocial/dissocial personalities, or uncontrollable outbursts, the intent to offend is a complex issue with significant implications for law enforcement. Generally, an individual's intent to offend is understood as (1) a rational decision based on a personal cost-benefit analysis of committing the criminal act or (2) a function of internal and external behaviors outside a person's control. The field of criminology has not come to a consensus on which perspective is correct, although there is a large research base testing the specific theories of each.

The rational choice perspective is probably the most applicable to this volume and should be interesting to defense policy makers because it treats an individual's intent to offend as constant. In this view, intent is more a function of how unprotected a target is than how capable the guardian is. The intersection of target vulnerability and guardianship in time and space presents an individual with the intent to offend with a nearly endless target set, assuming the individual has patience and the ability to conduct the cost/benefit (risk/reward) analysis. Law enforcement generally accepts this approach to offender intent, and myriad policies, procedures, technologies, and methods used in modern-day policing are based on this perspective.[5]

The general criminal investigative model that seeks to examine the way criminals perpetrate their offenses is also based on the assumption that the actor was rational in making decisions about when, how, where, and whom to victimize. This is generally referred to as modus operandi analysis and forms the basis for many of the pattern/signature matching approaches used to identify suspects in an investigation. The assumption that an offender's intent is based on rational choice allows investigators to reason about the facts/clues of an investigation. This reasoning can then be applied to other cases and modeled using information technologies like those mentioned previously to detect patterns of offending, to find matches in criminal behaviors (which can be thought of as criminal TTPs), and even to forecast future offending opportunities by modeling vulnerabilities and theorized guardian behaviors.

While previous patterns of behavior have not been conclusively proven to ensure accurate prediction of future behavior or intent, they do provide indicators and can contribute to analysis and moving the decision cycle "to the left," before intent becomes action. Inherent to this is a paradox of expertise

highlighted by Rob Johnston, based on the work of Ömer Akin, D. E. Egan, and B. J. Schwartz,[6] in which a distinction is drawn between a novice, who perceives randomness or disconnected data points, and a domain-specific expert, who sees patterns in the same data based on his or her organization of knowledge following exposure to and experience with thousands of cases. The paradox emerges in that strength can also be a weakness— experts are statistically superior to machines and novices in pattern recognition and problem solving based on their cumulative experience but less accurate at future predictions than Bayesian probabilities. Johnston adds, "An expert may know his [or her] specific domain, such as economics or leadership analysis quite thoroughly, but that may still not permit him [or her] to divine an adversary's intention, which the adversary himself [or herself] may not know."[7] The limitation on expert knowledge is proffered to manage expectations of any kind of panacea to discern intention.

The emergent recognition by elements of the Department of Defense of (1) the nexus between criminality and terrorism as a factor of force protection and (2) the value of examining terrorist modi operandi vis-à-vis TTPs has provided the opportunity to examine technologies which support the exploitation of this type of information.[8] One underemphasized contribution to this is the value of open-source intelligence, which often provides important strategic and operational insights into terrorist intent and capabilities. This is especially acute for the law enforcement community as there has been increasing pressure since the 9/11 attacks to provide national security clearances to additional law enforcement personnel only to have the process become mired in delay. The mainstay of policing organizations will remain in the open-source domain, and it is there that lessons relevant to war-fighter support can be learned. An example of a resource with demonstrated effectiveness is the Institute for the Study of Violent Groups (ISVG), which performs open-source research and exploitation on terrorist, extremist, and criminal organizations for the explicit purpose of enabling modus operandi analysis of activities. ISVG has assisted Special Operations Command, Pacific (SOCPAC) in assessing threats and intent posed by violent nonstate actor networks in Southeast Asia by examining networks and comparing modus operandi shifts in the activities of these networks. In several of the cases, the networks and components of support were unknown but knowable, and the illumination of each contributed to a more comprehensive product to enable both the analyst and the operator.

The results of the incorporation of open-source exploitation and the examination of intent through the lens of modus operandi analysis suggest, if not compel, a reconsideration of both tactics by the homeland defense and homeland security communities. Law enforcement has utilized the approach with demonstrated proficiency over time. Shoe-leather expertise is not meant to supplant other national technical means of collection and exploitation; it is rather a tremendous force multiplier in terms of knowledge discovery and time efficiency.

Notes

(All notes appear in shortened form. For full details, see the appropriate entry in the bibliography.)

1. In May 2009 the administration released the results of the 60-day cyber review and on 22 December of that year followed one of the key recommendations in appointing the nation's first cyber czar, Howard Schmidt.

2. *Streetcraft* is the operational art of law enforcement that is neither codified in any standard operating procedure nor taught in a police academy. Rather, it is learned on the street through the experience of dealing with the extremes of human behavior (as defined by Kiernan, "Hidden in Plain Sight").

3. See Rossmo, *Geographic Profiling.*

4. Applications of this include (1) geospatially modeling the physical and socioeconomic terrain of the US/Mexico border to better understand migrant smuggling routes and (2) geographically profiling the Washington, DC, sniper attacks in 2002.

5. The Compstat administrative approach for police resource allocation is based largely on a rational choice perspective of criminality. Many criminal investigative approaches, especially those that place an emphasis on examining criminal modi operandi, are founded on rational choice models.

6. Johnston, *Analytic Culture in the US Intelligence Community*; Akin, *Models of Architectural Knowledge*; and Egan and Schwartz, "Chunking in Recall of Symbolic Drawings."

7. Johnston, *Analytic Culture in the US Intelligence Community*, 66.

8. Early adopters include the Office of the Secretary of Defense and, in particular, the Rapid Reaction Technology Office, which evaluates and transitions capabilities to support the war fighter.

Bibliography

Akin, Ömer. *Models of Architectural Knowledge: An Information Processing View of Architectural Design.* London: Pion, 1980.

Coplink Web site. http://www.i2group.com/us.

Egan, D. E., and B. J. Schwartz. "Chunking in Recall of Symbolic Drawings." *Memory and Cognition* 7 (1979): 149–58.

Johnston, Rob. *Analytic Culture in the US Intelligence Community.* Washington, DC: Center for the Study of Intelligence, Central Intelligence Agency, 2005.

Kiernan, Kathleen. "Hidden in Plain Sight—Intelligence against Terrorism: Tradecraft or Streetcraft?" *Crime and Justice International*, July/August 2006, 29.

McKeithen, Katherine B., Judith S. Reitman, Henry H. Rueter, and Stephen C. Hirtle. "Knowledge Organization and Skill Differences in Computer Programmers." *Cognitive Psychology* 13, no. 3 (July 1981): 307–25.

Memex Technology Limited. Web site. http://www.memex.com.

Rossmo, Kim. *Geographic Profiling.* London: Taylor & Francis, 1999.

Tiburon. Web site. http://www.tibinc.com/solutions/lawenforcement.asp.

Tzu, Sun. *The Art of War.* Translated by Samuel Griffith. Oxford: Oxford University Press, 1971.

Chapter 2

Betting Responsibly

An Effects-Based Thinker's
Framework for Characterizing Intent

Col Harry A. Foster, USAF, Retired, Air War College

Abstract: This paper argues that effects-based thinking combined with a betting methodology offers the best analytic framework available to meld art with analytics to gauge intent. Adding betting methodology aids the analytic process by exposing the underlying presumptions and opinions. As the financial industry learned in the 1980s, quantifying human behavior with mathematical models is problematic. In the current decade, the effects-based operations initiatives fared little better. However, relationships do exist among nations, groups, and people. The perceptions and beliefs of nations, groups, and people do shape intent. Getting into the mind of another leader, therefore, requires an effects-based mental model to understand what political actions signal intent and what actions are merely noise. Assessing intent in a group context requires unconventional means to collect, analyze, and present the shared insights of the group.

As this volume establishes, understanding a state or non-state actor's intent is a squishy business. Recent attempts by the US military to quantify and measure an enemy's actions (including intent) through formal analytic systems like effects-based operations have been less than successful, with the commander of US Joint Forces Command stopping work on further development. Based on interviews with retired officers, some

15

best-selling authors have reinforced the notion that intent cannot be measured, suggesting instead that Clausewitz's notion of *coup d'oeil* or "gut feel" is probably the best we can do in this area.[1] For those who must sell their assessments to national-level decision makers, however, this approach is less than satisfying. The key question then is, are there mental or analytic models to help our gut?

This paper argues that effects-based thinking—separate and distinct from effects-based operations—combined with a betting methodology may offer the best analytic framework available to groups of planners trying to meld art with analytics to gauge intent. Since understanding intent for state and non-state actors is largely a *political study*, practicing effects-based thinking to gauge intent requires the strategist to ask only two questions: how are nation/group/person A and nation/group/person B related, and how does this relationship affect A's and B's actions? From this simple first-order analysis, the planner can build a larger mental map, which can then be used in concert with operational art to assess intent. Planners rarely work alone, however. Different planners tend to see the world differently. Adding a betting methodology to the analytic process aids the decision maker by exposing the underlying presumptions and range of opinions girding the analysis. To make the case, this paper begins by exploring the dimensions of intent. Next, it examines methodologies for gauging and measuring intent commonly used by military planners. Finally, the paper explores a betting methodology that employs effects-based thinking to characterize intent. To understand the methodology, one must first understand the dimensions of intent.

Background

Four decades ago, Thomas Schelling used the game of vehicular chicken as an analogy for geopolitics in his landmark work, *Arms and Influence*.[2] Thinking about the mentality behind a game of chicken is useful because it sheds light on the difficulty in understanding intent. The following briefly summarizes Schelling's main points:

- A game of chicken is, by definition, a game of nerve. Simply choosing to play or not to play is a choice unto itself and provides a clear signal of intent. For those that do play, the rules of the game are unclear because without uncertainty and unpredictability there is no game.
- Each player's perception of the other drives the way the game is played—a player who has a reputation for reck- lessness may be given different consideration than one who always yields.
- Third parties influence the game; therefore, the concept of saving face may be important. This outside pressure may force cooperation between the players as each tries to sig- nal his intent to play but not collide.
- Lastly, and perhaps most importantly, the options avail- able to each player narrow as the cars speed toward one another until reaching a point where neither player has a "last clear chance" to avoid a collision.

Although written during the height of the Cold War with nation-states in mind, Schelling's game-of-chicken analogy is still relevant at the strategic level, notwithstanding today's changed geopolitical environment with its mix of state and nonstate actors.

Using a single model to measure the intent of state and non-state actors may be controversial to some readers. Nonstate actors are organized very differently than governments and often employ asymmetric means in pursuit of their objectives. While they may differ organizationally and operationally from states, they are similar in that their overarching goals are political in nature with their operations carefully planned for maximum political effect, and, like states, their desired grand strategic ends are reasonably clear. Because of these similarities, a shared methodology for gauging intent is workable. There are limits to how far a politically based methodology extends within the nonstate domain, however. An ill-defined line separates the political nonstate actor from other groups such as organized crime. With these and Schelling's points in mind, I argue there are five broad dimensions of intent that the strategist must characterize to assess state and nonstate actors.

17

Dimensions of Intent

First, intent is multifaceted, shaped by a number of factors, including international and domestic perception of the other players, external and internal politics, and acts of third parties. Any model that gauges or measures intent, therefore, must attempt to characterize the impact, or *effect*, these interactions have on the players. Understanding who the key players are, what their relationship is, and how these relationships translate into influence is only one part of the equation, however. The strategist must also assess his or her nation's own network to understand how the opposing sides align and collide with one another. Understanding the historical and cultural drivers that shape *perceptions* within the networks is critical and discussed later.

Second, intent has bandwidth. That is, a player's intent is not a singular choice but represents a set of choices the player can make within a set of bounds or bookends. In the game of chicken, this bandwidth is defined by the sides of the road. In the real world, this bandwidth represents a set of bookends that define what an actor is *willing* to do, vice capable of doing. Unlike a road whose width never changes, the width between the bookends may change for each player as the game progresses. Characterizing this dimension of intent is the center of gravity in the planning process. The assumptions one makes in placing the bookends of both players, therefore, are critically important and represent significant risk. Accordingly, the strategist must constantly reassess the placement of these bookends, while policy makers continuously scrutinize underlying assumptions.

Third, intent is communicated though signaling. Signals may consist of overt or covert action, public or private diplomacy, media, or, importantly, inaction. The difficulty in signaling for both players lies in recognizing and correctly interpreting what is being communicated. As Graham Allison has pointed out, this is especially difficult given that organizations make decisions in at least three different ways: rationally, bureaucratically, or politically.[3] What may be viewed as an "irrational act" by one party may be viewed as completely logical by the other. Understanding this communication is made worse by the presence of incomplete and often conflicting information—Clausewitz's

18

notion of fog. Moreover, a player's true intent may be masked by tactics. Deception, trial balloons, or feints are as much a part of statecraft as they are of warfare. To deal with this environment, the strategist must develop methods that go beyond simple cost/benefit analysis to assess the actions of other players. Historical, cultural, religious, and organizational decision-making factors are important lenses and should be deeply integrated into analysis. Similarly, policy makers and strategists should consider how their actions are received and interpreted through the same lenses to ensure the signal intended is the signal received. As with assessing bandwidth, continually reviewing assumptions is critical.

Fourth, intent can become dynamic past a red line. Disputes can reach a point at which one or both players lose control of the situation. Beyond this red line, the bookends described earlier no longer apply. The fear that a conventional war in Europe would erupt into nuclear war was omnipresent during the Cold War. Today new concerns are emerging. Can a cyber attack attributed to a group cross a red line and result in military conflict between nations? Does an attack in space cross a red line and result in military conflict on Earth? Policy makers must choose where to draw these lines. They must also assess where the red lines of other players sit. Again, assumptions are critical as the stakes escalate quickly in these kinds of scenarios.

Finally, each player's own definition of success shapes intent. Absolute terms like "winning," "losing," or "achieved objectives" are not helpful in understanding this dimension. Instead, what shapes intent is perception—the degree of success or failure perceived by the network. For example, nonstate actors may consider an attack that decimates a number of their own cells a complete success if it generates positive coverage in the desired media and results in an increased number of recruits for its network elsewhere. As discussed earlier, this value system may force some to conclude an act was irrational. To the actor, however, the calculation is perfectly logical based on the perception of the relevant network. Accordingly, in estimating intent, the strategist should focus on characterizing the underlying value system that motivates the opposing actors vice struggling to understand rationality. Cataloguing the dimensions of intent is straightforward. Assessing it for decision makers is considerably harder.

Current Methods of Assessing Intent in Planning

To assess and measure intent, military planners using Joint Operation Planning and Execution System–type models traditionally form a specialized "red team" to study and characterize an opponent's strategic point of view. The team then represents that view throughout the planning process.

The red team characterizes intent and capability through development of "most likely" and "most dangerous" courses of action. The most likely course of action is largely a consensus assessment of intent—a prediction of how the enemy is expected to act given a set of political conditions and its collective capabilities. The most dangerous course of action is largely an assessment of capability—a prediction of how the enemy could act, given its collective capability, if political constraints are discounted and the situation spins out of control. The most likely course of action becomes the principal assumption of enemy behavior that drives planning for major operations. The most dangerous course of action is used to plan defensive measures and identify branch plans for consequence management and so forth.

In execution, intent is gauged by a set of priority intelligence requirements developed during planning. Derived from the most likely or most dangerous enemy courses of action, these requirements tell intelligence collectors where to look and assessors what to look for. Underlying each of these intelligence requirements may be a set of warnings and indicators that signal intent: increasing readiness, allocations of resources, or movement of forces. Measuring intent in execution is also subjective, but it takes into account enemy action or inaction through objective indicators like warnings and indicators.

There are few formulaic models or templates to assist team leaders in guiding a red team through an assessment of red intent. Students of strategy may reach for Neustadt and May's *Thinking in Time* for ways to frame the issues or draw on Allison and Zelikow's *Essence of Decision* to think about how different types of organizational behaviors influence outcomes. Beyond these texts, planners draw on experience gained through education and personal study. For those who buy into the *coup d'oeil* school of characterizing intent, the lack of analytic models to characterize intent may be viewed as a good thing. However, given the realities of team size and time pressures, a general framework for gauging intent is useful even if it is used only as a starting point.

A Betting Methodology for Characterizing Intent

The premise of this effects-based framework is that intent is fundamentally a political calculation. While the focus of one nation's intent may be aimed at the other player in a game of chicken, intent is shaped by a number of factors (e.g., culture, history, third parties, domestic polity, governmental power structures), not just the other player. If the strategist can develop some idea of relationships in this influence structure, then a player's true intent may be easier to "bookend" through analysis. Accordingly, the following offers a brief planning framework which uses a betting system to make an effects-based characterization of intent.

Step 1: Set a Common Frame of Reference

Any understanding of intent begins with some degree of understanding of the cultural history and the history of the issue in question. This may be obvious, but it is hard to enforce in execution. Each person who judges a situation comes into it with his or her own mental model of how the world works. In an effort to simplify and understand, the natural tendency is for people to rely on analogies based on their own experience. Analogies, however, can lead to the wrong conclusions when dealing with different cultures, forms of government, and worldviews.[4] Therefore, everyone's mental model—regardless of rank, background, and experience—needs calibrating. This can be accomplished through reading or briefing. The key point is that some amount of time must be spent getting into the mind of the red nation before attempting to assess its actions. Leveling background information is not enough, however.

Setting a common frame of reference also requires developing a method to address analogies. Practices such as banning the use of analogies among planners do not work—they remain in the mind, unspoken. May and Neustadt offer a simple method to address the problems analogies raise: for any given analogy, ask "what's like this case?" and "what's different?" This technique offers a simple, quick way to reset one's mental model to the task at hand.[5] Building on this idea, planning team leaders should record the analogies red team members raise, perform like/different analysis, and add it to the background brief. When combined with the cultural background, addressing analogies up front is a powerful tool in bringing team members' minds into adjusted focus.

Step 2: Describe the Issue at Hand

After leveling the background, the next step is to describe the issue at hand *from each side's point of view*. Defining the issue up front is important for a couple of reasons. First, it is vitally important to capture what's known (and how), what's presumed (and why), and what's unknown. Information will never be perfect, so making assumptions is a necessary part of practicing operational art. However, a big part of measuring intent (discussed later) depends upon how solid the information underlying the analysis is. Second, defining the issue early allows the team members to challenge it often throughout the process. Perceived irrational behavior by a player probably means the team has the issue at hand wrong.

Step 3: Map the Players and Their Relationships

Mapping the players and their relationships is the core element in effects-based analysis. While there may be many elements one could map, the principal elements of interest here are ones that directly affect the decision of each nation. As relationships are mapped, the strategist should assess the relative degree of influence these connections represent.

Internal to each nation is a set of relationships that defines its decision-making process. This may consist of constitutional structures, constituencies, and bureaucracies in democratic governments or influential elites, parties, and committees in more centrally managed governments or among nonstate actors.

External to each nation are connections with other nations or nonstate groups that can affect a state or nonstate actor's decision process. These connections may include cultural, religious, diplomatic, economic, or military relationships. These relationships are important not only because they offer insight into possible sources of leverage, but also because they may offer an avenue for signaling between the parties.

Step 4: Define the Main Player's Strategic Goals and Red Lines (Assess Rationally Based Decision Making)

Once relationships are mapped, the planning team should assess each player's strategic goals and possible red lines with regard to the issue at hand. This gain/loss-based analysis allows the strategist to consider each side's decision calculus from a rationally driven decision-making approach. The purpose of this analysis is twofold: to examine how the issue at hand fits into the larger context of the nation's strategic goals and to explore factors that may serve to restrain the actions of each side such as red lines.

Step 5: Define Where Third Parties and Internal Parties Stand (Assess Politically Based Decision Making)

To consider the effect of a politically driven decision-making approach, the planning team should map where third parties and internal parties stand, if possible. The purpose of this analysis is to understand how groups around the decision maker may drive intent, despite the possible setbacks to their strategic goals.

Step 6: Assess Available Choices and Likely Indicators

At this point, it is time to assess intent itself. In keeping with the concept that intent is not a singular choice, but represents a set of choices bounded by bookends, each red team member offers his or her assessment of intent. This consists of a set of plausible actions that define the bookends and the physical

23

indicators or signals that provide verification. These choices are then aggregated into a list for further analysis. Indicators are compiled into a list of prioritized intelligence requirements to drive intelligence collection.

Step 7: Measure Intent by Taking Bets

The final step in an effects-based framework for character-izing intent is measurement. As this paper has established, measuring human behavior is difficult on its face. However, in planning environments, staff culture complicates it even more. First, military culture often demands that staffs present their boss with "the answer"—a certain assessment of what will hap-pen. This can force hesitation as planners wait for confirmed data. New data may invalidate parts of the old assessment, driving requirements for more data and more assessment. Waiting for "good" data can literally cripple the effort. Second, in an effort to preserve a senior decision maker's time and pre-vent information overload, the cognitive style of staff processes may limit assessments of intent to bullet points on two or three PowerPoint slides. As a result, the range of staff views is often suppressed, and the subjective feel of the assessments is lost as analysis is simplified or placed in backup. As Neustadt and May point out, this is a reality of government.[6]

An alternative that overcomes both of these staff-driven com-plications is to set up a betting system.[7] A notional betting system that assesses intent consists of three parts. In part one, assessors rate their opinion of the quality of the information underlying their assessment on a scale of one to ten. Plotting aggregated high, low, and mean data on a Gantt chart provides senior decision makers with a visual, qualitative summary of the information underlying the overall assessment. In part two, assessors rank order the top five key relationships that they feel most influenced the intent of each player. This assessment may help senior decision makers key in on critical details in a flood of intelligence data to inform their own mental models. In part three, assessors place bets against the list of available choices based on the probability that the move represents an opposing nation's intent. When graphed on a bar chart, this assessment not only points the senior decision maker to the

most likely intent but also provides a visual depiction of the range and strength of opinion regarding other possibilities. Assessors bet at set intervals, say, every 48 hours. Regular intervals allow senior decision makers to get a quick sense of the speed and direction of changes in the staff's assessment as information and assumptions change.

Step 8: Present the Data

The best analyses in the world are irrelevant if the message is not received by the decision maker. Assessments of intent should include a list of key facts, assumptions, and unknowns (with changes annotated); a sketch outlining the relationships analyzed along with the team's assessment of the top five relationships; a Gantt chart showing the team's confidence in the underlying data (with maximum, minimum, and mean); and a bar chart showing the distribution of bets for each possible action. Assessment to assessment, trend data is important. Presenters should highlight changes in the team's underlying data over time and changes in betting over time. In addition, presenters should highlight key disagreements between groups when more than one group bets.

Conclusion

As the financial industry learned in the 1980s, quantifying human behavior with mathematical models is problematic. In the current decade, the effects-based operations initiatives fared little better. However, there is no denying that relationships do exist among nations, groups, and people. Their perceptions and beliefs do shape intent. Getting into the mind of another leader, therefore, requires an effects-based mental model to understand what political actions signal intent and what actions are merely noise. Assessing intent in a group context requires unconventional means to collect, analyze, and present the shared insights of the group.

Notes

(All notes appear in shortened form. For full details, see the appropriate entry in the bibliography.)

1. Clausewitz, *On War*; and Gladwell, *Blink*, 99–147.
2. Schelling, *Arms and Influence*, 116–25.
3. Allison and Zelikow, *Essence of Decision*, 3–11.
4. Neustadt and May, *Thinking in Time*, 75–90.
5. Ibid., 89.
6. Ibid., 1.
7. Ibid., 159.

Bibliography

Allison, Graham, and Philip Zelikow. *Essence of Decision: Explaining the Cuban Missile Crisis*. 2nd ed. New York: Longman, 1999.

Clausewitz, Karl von. *On War*. Translated by O. J. Matthijs Jolles. New York: Modern Library, 1943.

Gladwell, Malcolm. *Blink: The Power of Thinking without Thinking*. New York: Little, Brown and Co., 2005.

Neustadt, Richard, and Ernest May. *Thinking in Time: The Uses of History for Decision Makers*. New York: Free Press, 1988.

Schelling, Thomas. *Arms and Influence*. New Haven, CT: Yale University Press, 1966.

Chapter 3

Gauging the Intent of Nation-States and Nonstate Actors

An Operator's Perspective

Gary Schaub, Jr., PhD, Air War College

Abstract: How should policy makers approach divining the intentions of adversaries who may take actions that the United States wishes to deter? Determining adversarial intent during the Cold War was based upon capabilities analysis married to worst-case scenarios of what the adversary could accomplish. The *Deterrence Operations Joint Operating Concept* (*DO JOC*) revised this thinking by recognizing that an adversary has a choice between complying with a demand to refrain from action and defying that demand—and that the adversary will consider the expected value of each of these options. This has opened significant doors to making the deterrence planning and assessment processes used by the US military, from Strategic Command to the regional combatant commands, much more sophisticated and, hopefully, effective.

Currently, however, there is no process or framework to help analysts determine adversarial intent. A process needs to be established to infer adversary intent on a continuous basis so that a usable product is available to assist in routine planning or in the event of a crisis. One method of achieving this goal is competing hypotheses that encourage a debate and critical thinking about the adversary to prevent typical intelligence errors such as mirror imaging or groupthink. This intent-assessment process would allow debate and discussion that could inform a commander or political leader about the issues, foreign and domestic, that are pressing on the adversary's leadership, provide his or her planning staff the basis for recommending whether deterrence or some other strategy is wise in the present circumstances, and also provide a basis upon which to assess the likelihood of success.

Introduction

The *Deterrence Operations Joint Operating Concept* (*DO JOC*) defines deterrence operations as those that "convince adversaries not to take actions that threaten US vital interests by means of decisive influence over their decision-making. Decisive influence is achieved by credibly threatening to deny benefits and/or impose costs [if the undesirable action is taken], while encouraging restraint by convincing the actor that restraint will result in an acceptable outcome."[1] The *DO JOC* thus takes an active view of deterrence operations: achieving decisive influence over an adversary's decision making requires deliberate action on the part of a joint force commander or other American policy makers. Such deterrence operations can include force projection, the deployment of active and passive defenses, global strike (nuclear, conventional, and nonkinetic), and strategic communication.

The key to knowing when to practice deterrence is determining an actor's intent. Patrick Morgan notes that "the intentions of opponents are notoriously difficult to fathom."[2] How do joint force commanders, those who populate the staffs of the US government, and the elites upon whom they rely for subject matter expertise determine adversary intent? Do military staffs rely on doctrinal guidance to perform this key task? Are certain patterns of thought or interpretive lenses commonly employed by officers, civilian policy makers, or scholars? How have these been applied in key episodes in the past? Finally, how can the process of intent determination be improved?

Doctrinal Guidance

There is little doctrinal guidance for determining adversary intent. What exists is contained in Joint Publication (JP) 2-0, *Joint Intelligence*. This doctrine manual contains superficially useful sections, such as "Intelligence and the Levels of War," "Intelligence and the Range of Military Operations," "Prediction — (Accept the Risk of Predicting Adversary Intentions)," and "Intelligence Support during the Deterrence Phase." Unfortunately, most of these sections are unhelpful.

Beyond exhorting "intelligence professionals" to "go beyond the identification of capabilities" and take the risk of predicting adversary intent, basing such forecasts on "solid analysis," JP 2-0 is not particularly helpful in guiding such analysis. Indeed, by indicating that such "an intelligence product . . . usually reflects enemy capabilities and vulnerabilities," the authors of this doctrine indirectly encourage that capability analysis be substituted for intent analysis. While capabilities do suggest some general directions of intent—why invest in a particular capability if you are not going to use it?—capability analysis utterly fails to answer questions of the conditions under which such capabilities would be used. These are political issues that the military intelligence process, set as it is at the tactical or operational level of war, does not address.

Interpreting Intent: Two Frameworks

If joint military doctrine is not a helpful guide in determining adversary intent, how can operators structure this problem so as to solve it? Intelligence analysts operate in a complex environment, and they, like all human beings, are unable to process all of the innumerable stimuli that they encounter. In this context, Roberta Wohlstetter usefully distinguished "between signals and noise."[3] What Wohlstetter left unsaid was that noise and signals do not come clearly marked for the analysts as they sift through mountains of information. Rather, it is the analyst who determines what is signal and what is noise.

This is a difficult task. Analysts suffer the same cognitive limits as everyone else and therefore necessarily deal with "a dramatically simplified model of the buzzing, blooming confusion that constitutes the real world."[4] These simplified models of reality focus one's attention toward certain pieces of information and away from most others and generally represent the "most significant chains of causes and consequences" as "short and simple."[5] These models allow analysts to discriminate between signals and noise. Consequently, it is up to the analyst to determine which information best explains the adversary's intent.

American scholars and policy makers have been apt to apply one of two models to comprehend the intentions of other international actors, be they states or nonstate organizations engaging in politics: the strategic intent model and the internal logic model. Each model posits that the actor is purposive—the actor seeks to achieve a particular goal with each action. When the analyst is working retrospectively, this presumption risks making either framework tautological, as "an imaginative analyst can construct an account of value-maximizing choice for any action or set of actions."[6] This tautology can be escaped, however, if one also presumes that the preferences against which alternatives are considered are relatively stable. This allows an analyst to erect a set of principles that appear to guide the actor's choices over time and across domains. These principles fill in generic references to preferences or utilities for particular actors and allow some degree of operationalization of the model. The preferences and utilities can be derived from "(1) propensities or personality traits or psychological tendencies of the nation or government [or nonstate organization], (2) values shared by the nation or government [or organization], or (3) special principles of action [that] change the 'goals' or narrow the 'alternatives' and 'consequences' considered."[7]

The strategic intent and internal logic models differ with regard to the problems that they believe an actor is attempting to solve by taking actions in the interstate arena. The strategic intent model presumes that state and nonstate actors direct their behavior toward achieving political goals vis-à-vis external actors. It presumes that they desire to influence the decisions, behavior, and/or attitudes of these external actors and that they have chosen the most effective means available to them, as delimited by their capabilities and tendencies, to achieve this end. Whether they do so via coercion, inducement, or persuasion,[8] using whatever power resources they have available, matters not. What does matter is that the impact on the external actor is of paramount concern to the adversary.

Thus the key variables determining the adversary's intent to act are the costs of undertaking the action, the benefits that would accrue from successful action, and the costs and benefits of not acting. The strategic intent model is vague with regard to what factors determine costs and benefits of these two courses of action. Lawrence Freedman has argued that the costs of undertaking the action can be bifurcated into those costs associated with implementing the choice and those associated with enforcing it after the fact.[9] The benefits of undertaking the action have not been given as much attention as the costs but would be composed of material benefits accrued, intangible benefits—including prestige, reputation, and so on—and the new opportunities made possible by successful conclusion of the action. The costs of inaction, or "restraint" in the parlance of the *DO JOC*, can be broken down into the international and domestic costs of forgoing action, including suffering the unwanted reactions of opponents in the near and far term and the negative reactions of domestic audiences. The benefits of inaction or restraint have not been well thought out in the literature either but would include desirable international and domestic reactions—such as praise for being reasonable or a de-escalation of tensions, or tangible benefits provided by those who did not favor action. Despite the obvious utility of considering domestic reactions to the choice made by the adversary's leadership, the strategic intent model generally focuses upon externally generated costs and benefits.[10]

The internal logic model, on the other hand, presumes that actors are directing their activities inward, pursuing advancement or preservation of the group, and that actions directed toward other actors—be they states or otherwise—are judged primarily by their internal effects rather than their external effects. Hence international political behavior is primarily a consequence of domestic (or internal) politics and may be more incidental than intended. "The idea that political elites often embark on adventurous foreign policies or even resort to war in order to distract popular attention away from internal social or economic problems and consolidate their own domestic political support is an old theme in the literature on international politics," argues Jack Levy.[11] Ned Lebow argues that states with weakening political systems, weakening political leaders, or elites engaged in a competition for power may fall back on "the time-honored technique of attempting to offset discontent at home by diplomatic success abroad."[12] While success vis-à-vis external actors would certainly be welcomed, the cohesion within the group and support for the leadership generated by conflict abroad are the primary purposes of such actions.

The key variables within this framework are the internal or domestic groups whose support is required for the continued functioning of the state or nonstate organization. After they have been identified, the relative ability of these groups to influence the leadership by providing benefits such as continued support or imposing costs such as removing the leadership from power, the audiences' views of the merits of the action to be undertaken (or not), and the relative ability of the leadership to substitute the support of one group for another must be assessed.[13] Thus the internal logic framework requires substantial knowledge of the adversary beyond the leadership and its preferences. It requires detailed knowledge of the domestic political situation if the adversary is a state or of the internal dynamics of a nonstate organization. A substantial body of work has addressed the propensities of certain types of regimes to engage in external behavior to ameliorate internal dissension or promote internal cohesion, democratic states in particular.[14] The manner in which deterrent threats are interpreted and used when internal needs drive external behavior has received attention from scholars such as Ned Lebow and Janice Stein, but their insights have not been incorporated into the corpus of deterrence theory.[15]

American policy makers, scholars, and analysts have relied upon these two frameworks of rational action to infer the intent of adversaries. They clearly direct attention toward different aspects of the adversary's makeup, his capabilities, and particularly the hierarchy of his goals. Unsurprisingly, they often provide contradictory prescriptions with regard to how to approach an adversary and what to do to influence his behavior. A short example of each model in action should make their differences clear.

Terrorist Objectives

In the immediate aftermath of 9/11, the public, the media, and some policy makers tended to eschew either model of rational and purposive adversary behavior in favor of an instinctual one, which posited that Islamic terrorists such as those in al-Qaeda "hate us for who we are rather than what we do."[16] Similar language was included in the 2002 *National Security Strategy of the United States*, which identified "rogue states" as those that "reject basic human values and hate the United

States and everything for which it stands."[17] When one posits that adversary intent derives from raw emotion such as hatred and such emotion permeates all members and aspects of an adversary's organization—be it a state or a nonstate actor— strategic thought is likely to be bypassed in favor of brute force.

In the analytic community, however, affective models of adversary behavior have not been paramount. Indeed, the strategic intent model has been primary.[18] As Max Abrahms put it,

> The strategic model assumes that terrorists are motivated by relatively stable and consistent political goals. . . . Second, the strategic model assumes that terrorism is a "calculated course of action" and that . . . terrorist groups weigh their political options and resort to terrorism only after determining that alternative political avenues are blocked [or at least not as efficacious], . . . [and] they possess "reasonable expectations" of the political consequences of using terrorism based on its prior record of coercive effectiveness.[19]

The strategic intent model also applies to suicide terrorism, for which motives have often been identified as religious fanaticism or insanity. As Bob Pape argues, "What nearly all suicide terrorist attacks have in common is a specific secular and strategic goal: to compel modern democracies to withdraw military forces from territory that the terrorists consider to be their homeland."[20]

This has been reflected in policy framing as well. As Pres. George W. Bush put it in his address to Congress on 20 September 2001, "Al Qaeda is to terror what the Mafia is to crime. But its goal is not making money, its goal is remaking the world and imposing its radical beliefs on people everywhere. . . . These terrorists kill not merely to end lives, but to disrupt and end a way of life. With every atrocity, they hope that America grows fearful, retreating from the world and forsaking our friends. They stand against us because we stand in their way."[21]

Prescriptions that are derived from the strategic intent framework suggest that terrorists can be deterred by increasing the difficulty of their efforts to execute their strategy or by imposing costs on the groups involved through sanctions or other forms of punishment. These prescriptions also suggest that terrorists can be placated by concessions that allow them to achieve many of their objectives without resorting to violence.[22]

As Abrahms puts it, these "are designed to reduce terrorism by divesting it of its political utility."[23] Over time, these analysts argue, as the terrorists' strategy of coercion is both frustrated tactically and successful strategically, they will moderate their behavior and be co-opted into the normal political processes of the state—be it their own, as happened with the Palestinian Liberation Organization, or that of their former adversary, as happened with the Irish Republican Army.

On the other hand, the internal logic model has also been utilized to explain terrorism. Paul Davis and Brian Jenkins have argued that "deterrence [of terrorist groups] is . . . difficult because for many of the people involved, terrorism is a way of life. . . . Terrorism provides 'positives'—notably status, power, recruits, and psychological rewards."[24] Mia Bloom argues that "under conditions of mounting public support, [suicide] bombings have become a method of recruitment for militant Islamic organizations within the Palestinian community. They serve at one and the same time to attack the hated enemy (Israel) and to give legitimacy to outlier militant groups who compete with the Palestinian Authority for leadership of the community."[25] Bloom further argues that as the intifada continued and Yasser Arafat's Palestinian Authority lost its monopoly over the legitimate use of force—legitimate in the eyes of the Palestinian people— "groups competed and outbid each other with more spectacular bombing operations and competition over claiming responsibility. At the same time, the operations whipped up nationalist fervor and swelled the ranks of Islamic Jihad and Hamas, who used the bombings, in conjunction with the provision of social services, to win the hearts and minds of the Palestinians."[26]

The use of terror operations in the competition among these groups for leadership of the movement and recruitment and retention of members is to the detriment of their strategic cause, argues Bloom,[27] and has left Palestinians worse off than they were before the suicide bombing campaigns began. Andrew Kydd and Barbara Walter argue, as does Abrahms, that Palestinian terrorists prefer to continue their activities in spite of the possibility of achieving their political goals through less violent means—or even as a result of successful coercion.[28] They therefore act as spoilers to any political settlement and perpetuate the conflict that provides their raison d'être. The violence is not a means to a political end vis-à-vis their adversary but, instead, a means to achieve a sense of honor, group worth, and identity.[29] Indeed, the effects of violence in these areas have even been termed a "public good" for the group by one terrorism analyst.[30]

The internal logic framework suggests that the internal dynamics of terrorist groups drive their activities, not the potential attainment of a strategic goal. This suggests that influencing their behavior will be difficult absent destruction of the terrorist groups and those that support them. Indeed, Abrahms argues that "strategies to dry up demand for terrorism by minimizing its political utility are misguided and hence unlikely to work."[31] The October 2002 *National Security Strategy of the United States* argued that "traditional concepts of deterrence will not work against a terrorist enemy."[32] Because of this, President Bush argued that "our security will require all Americans to be forward-looking and resolute, to be ready for preemptive action when necessary."[33] To many, this leaves brute force to eliminate the adversary as the only effective policy.[34] As Ralph Peters put it, "Until a better methodology is discovered, killing is a good interim solution."[35]

Prescriptive Problems

The strategic intent model and the internal logic model of adversary intent produce very different pictures of what motivates the adversary. Does he desire to influence external actors so as to achieve a political outcome vis-à-vis that actor? Or does he desire to bolster the solidarity of his group in the face of centripetal forces? Is the outcome of the action that we wish to deter of primary or secondary importance to the adversary? Making this determination is important when deciding whether to attempt to deter the adversary's actions or to take another approach, such as preemptive brute force or actions to increase or decrease the adversary's feelings of insecurity.

Deterrence is a strategy to pursue when one judges that the adversary's resort to arms is motivated primarily by strategic goals. Given that it is directed toward external actors in such situations, identification of the adversary's goal is a matter of routine. Focusing deterrent demands toward that objective—

"don't do that"—places the adversary in a decision situation in which he can either comply with what has been demanded of him or defy those demands and risk the implementation of the deterrer's threatened sanction. As the *DO JOC* rightly suggests, denying the adversary the potential benefits of the actions that he intends to take or imposing costs that reduce the net utility of the action are the two ideal ways of reducing the likelihood that the adversary will choose to act. The objective of this deterrent threat is to reduce the expected value of "doing that" to a point that the consequences of compliance are of greater value. As the *DO JOC* explains, "Adversaries weigh the perceived benefits and costs of a given course of action *in the context of* their perceived consequences of restraint or inaction. Thus deterrence can fail even when the adversary perceives the costs of acting as outweighing the benefits of acting if he believes the costs of inaction are even higher still" (emphasis added).[36] When the adversary is basing his choice upon these considerations, deterrence is correctly targeted and has a chance of success.

Deterrence may not be the strategy to pursue if the adversary's external behavior is directed toward enhancing internal cohesion or the power of the leadership. Providing overt signs of an external threat is precisely the outcome desired by the adversary's leadership. This external threat allows the leaders to take actions to increase their support, silence moderates or critics, mobilize resources that might otherwise be unavailable, and provide the opportunity for common identities to be forged or reinforced.

If the adversary is motivated by internal logic, is it really a no-win situation for the deterrer? Or are there alternatives to issuing an immediate deterrent threat directed against the adversary's intended external action or doing nothing and letting the adversary's provocation pass unanswered? There are a number of options.

First, one can still attempt to deter the adversary directly through passive measures that deny him the opportunity to carry out his aggressive intent and also deny him the visible indicators of hostility that he seeks to engender. A number of means can be used to do this. One denial measure is to harden soft targets—be they intercontinental ballistic missile (ICBM) silos or police stations—through passive point defenses. These defenses make it less likely that spectacular successes can be had against these targets, and given their passivity—barriers, reinforced concrete, and so on—they deprive the deterrer of the ability to overreact and justify the adversary's actions.[37] Passive area defenses can also be used to deny the adversary the interaction he needs with the deterrer to achieve his internal goals. Possibilities in this realm include measures such as the fence that Israel erected around Palestinian areas, which has decreased suicide attacks substantially since its completion,[38] or diplomatic isolation such as that imposed upon the People's Republic of China, Cuba, and Iran after their revolutions. A potential drawback to passive area defenses is that they themselves might become symbols of implacable and unyielding hostility that the adversary can use repeatedly to rally its domestic constituents.[39]

Second, one can attempt to deter the adversary indirectly—by directing the deterrent threat toward the members of the group that the leadership is attempting to bolster or recruit from. The adversary's external challenge is designed to attract these followers, and a deterrent threat that is directed toward the group's members and potential members may cleave them away by highlighting personal over group interests.[40] All groups engaged in conflict who are attempting to recruit or retain members ask these people to put aside their personal interests for the benefit of the group's cause, even though their individual contributions will be marginal (in most cases—suicide terrorism is designed to overcome this recruitment challenge). "Thus rebels confront the possibility of disastrous private costs and uncertain public benefits. . . . Unless the collective action problem is somehow overcome, rational people will never rebel—rebellions, that is, require irrationality."[41] Israel has pursued such a deterrent policy by threatening to destroy the family homes of young Palestinians who were involved in attacks. Such an option would be an attempt to deny the adversary leadership the domestic benefits of his intended action by threatening to punish individual members of the group.

Third, one can pursue a similar goal through inducements to members of the adversary's constituency rather than through coercion. Counterinsurgency (COIN) strategies, such as those discussed in Field Manual (FM) 3-24, *Counterinsurgency*, work on this principle: "The real battle is for civilian support for, or acquiescence to, the counterinsurgents and host nation government. The population waits to be convinced. Who will help them more, hurt them less, stay the longest, and earn their trust?"[42] Indeed, the Anbar Awakening in Iraq is quite a vivid example of using inducements to cleave potential supporters away from an adversary—in this case al-Qaeda in Iraq.[43]

Fourth, one can attempt to "encourage adversary restraint," as the *DO JOC* puts it, by "try[ing] to communicate . . . benign intentions . . . to reduce the fear, misunderstanding, and insecurity that are often responsible for unintended escalation to war."[44] Engaging in such persuasion is an alternative to influence through coercion or inducement. It involves altering the considerations by which compliance and defiance are evaluated. The persuader does not promise or threaten action but convinces the adversary to see the situation in such a way that he realizes it is in his own interests to act differently. This can be done by highlighting—without altering—costs or benefits related to complying with or defying the persuader's demands or by offering new alternatives that allow the adversary to achieve his goals in ways that do not harm the persuader's interests. These persuasion strategies treat the definition of the problem facing the adversary—in this case increasing cohesion, recruitment, or retention of members—as given or settled. Another avenue of persuasion requires understanding the basis upon which the target frames the issue.[45] Persuasion is generally seen as a fruitless option, particularly when dealing with an adversary whose primary concerns are internally generated, although this judgment may have more to do with the willingness to engage the adversary in terms that provide legitimacy than with an objective assessment of the chances for success.

Fifth, one can forgo influence altogether and use brute force against the adversary to prevent him from undertaking action.[46] This can take the form of disarming the adversary to deny him the capability to pursue the action that he intends or decapitating the adversary so as to disrupt his ability to act. Either action risks increasing the cohesion of the adversary by justifying his hostility toward the deterrer and/or creating a martyr of the leadership. Decapitation of the leadership could also disrupt the internal cohesion of the adversary to some degree.[47]

Overall, if it is determined that an adversary decision maker is motivated by the internal logic of his group's situation, deterrence may work—but not in the manner prescribed in the *DO JOC*. Rather, deterrent demands and other influence attempts should be directed at the primary objectives of the adversary in these situations: the internal constituencies whose support he hopes to rally by his external actions. Clearly, measures should also be taken to mitigate the impact of those actions, since nothing fails like failure. But mere signals of hostility directed toward the group (or nation) as a whole in an attempt to deter the unwanted action could provide the adversary leader precisely what he wants: an external enemy that his people can oppose in unity.

Conclusion

How should policy makers approach divining the intentions of adversaries who may take actions that the United States wishes to deter? Although deterrence formed the core mission of the American military throughout the Cold War,[48] adversary intent was based upon capabilities analysis married to worst-case scenarios of what the adversary could accomplish. Whether deterrence would succeed in general or in any particular case was likewise inferred to be a function of American capabilities and willingness to use them in the event that deterrence failed. What would happen if deterrence succeeded and the adversary's intent was frustrated was rarely considered.

The *DO JOC* rectified a basic problem in previous deterrence thinking by recognizing that an adversary has a choice between complying with a demand to refrain from action and defying that demand—and that the adversary will consider the expected value of each of these options. No longer is "restraint" considered to be an option that is outside of the deterrence calculus for the adversary or the deterrer. This has opened significant doors to making the deterrence planning and assessment processes used by the US military, from Strategic Command (STRATCOM) to the combatant commands (COCOM), much more sophisticated and, we hope, effective.

Getting the basic framework correct has led to the next issue: determining how much the adversary desires to undertake particular actions, those the United States would prefer that the adversary not undertake, and others that might provide less offensive alternatives. This requires assessing adversary intent. Regrettably, there is no set process or framework for undertaking this necessary analysis. JP 2-0, *Joint Intelligence*, merely exhorts intelligence analysts to "take risks" to "predict" adversary intent. Intelligence officers, uniformed and civilian, have indicated that producing such analyses is considered more of an art than a science and that no processes have been established; rather, intelligence analysts are left to develop their own methods to produce their analytic products. Hoping that particular analysts in key positions are da Vincis or Michelangelos is simply unacceptable. Military staffs excel at planning and use set processes to yield acceptable and improvable products. Such a process needs to be established to infer adversary intent on a continuous basis so that a usable product is available to assist in routine planning or in the event of a crisis.

Such a process should begin with a skeleton framework that focuses on producing at least two narratives of adversary behavior: a strategic intent model and an internal logic model. As I have discussed in the preceding sections, these two frameworks have provided the basis for rival interpretations of adversary behavior such as that of the Soviet Union during the Cold War and terrorist organizations today. They have also provided alternative prescriptions for American behavior. Their explicit use would allow debate and discussion in the intent-assessment process that could inform a commander or political leader about the issues, foreign and domestic, that are pressing on the adversary's leadership, provide his or her planning staff the basis for recommending whether deterrence or some other strategy is wise in the present circumstances, and also provide a basis upon which to assess the likelihood of success. Developing an intent-assessment process would also help to operationalize and institutionalize the Department of Defense's current concerns with cultural competency and provide the basis for the personnel system to reward those officers who excel in this particularly useful but heretofore neglected area of professional expertise. Thus many goods would follow from a more coherent and systematic process of assessing adversary intent.

Notes

(All notes appear in shortened form. For full details, see the appropriate entry in the bibliography.)

1. US Strategic Command, *DO JOC*, 8.
2. Morgan, *Deterrence Now*, 30.
3. Wohlstetter, "Cuba and Pearl Harbor," 691.
4. Simon, *Administrative Behavior*, 3rd ed., xxix.
5. Ibid., 4th ed., 119.
6. Allison, *Essence of Decision*, 35.
7. Ibid., 36–37.
8. For the relations among these forms of influence, see Joint Forces Command, *Strategic Communication Joint Integrating Concept*, 101–3.
9. Freedman, "Strategic Coercion," xx.
10. Most empirical studies of deterrence exclude domestic-level independent variables from their analysis. See Schaub, *Deterrence, Compellence, and Rational Decision Making.*
11. Levy, "Diversionary Theory of War," 259.
12. Lebow, *Between Peace and War*, 66–69.
13. For an overly abstract discussion of these variables, see Bueno de Mesquita, *Logic of Political Survival.*
14. For a recent study, see Brulé, "Congress, Presidential Approval, and U.S. Dispute Initiation."
15. Rather, their work is seen as providing an alternative to deterrence theory. See Lebow and Stein, "Beyond Deterrence"; Lebow and Stein, "Rational Deterrence Theory"; Huth and Russett, "Testing Deterrence Theory"; and Lebow and Stein, "Deterrence: The Elusive Dependent Variable."
16. Hanson, "Islamicists Hate Us for Who We Are, Not What We Do," 1.
17. *National Security Strategy*, 15.
18. Early work includes Crenshaw, "Logic of Terrorism"; and Schelling, "What Purpose Can 'International Terrorism' Serve?"
19. Abrahms, "What Terrorists Really Want," 80–81.
20. Pape, *Dying to Win*, 4.
21. Bush, Address to Congress, 20 September 2001.
22. Byman, "Decision to Begin Talks with Terrorists"; and Neumann, "Negotiating with Terrorists." Jones and Libicki found that 43 percent of those terrorist groups that ceased to exist between 1968 and 2006 did so because their goals were accommodated by political authorities. See Jones and Libicki, *How Terrorist Groups End.*
23. Abrahms, "What Terrorists Really Want," 103.
24. Davis and Jenkins, *Deterrence and Influence in Counterterrorism*, 5.
25. Bloom, "Palestinian Suicide Bombing," 61.

41

26. Ibid., 71.

27. Bloom, "Palestinian Suicide Bombing," 71.

28. Kydd and Walter, "Sabotaging the Peace"; and Abrahms, "What Terrorists Really Want."

29. Bloom, "Palestinian Suicide Bombing," 74.

30. Argo, "Banality of Evil," cited in Bloom, "Palestinian Suicide Bombing," 13.

31. Abrahms, "What Terrorists Really Want," 103.

32. *National Security Strategy*, 15.

33. Bush, graduation speech.

34. Alexander, "International Relations Theory Meets World Politics," 19.

35. Peters, "Thou Shalt Kill," 19.

36. US Strategic Command, *DO JOC*, 26–27.

37. This is an application of Schelling's art of the commitment. See Schelling, *Strategy of Conflict*, 21–52.

38. Byman, "Do Targeted Killings Work?" 105–6.

39. Kalman, "Israeli Fence Puts 'Cage' on Villagers."

40. Tolbert, *Crony Attack*; and Lichbach, *Rebel's Dilemma*. On the other hand, Jerrold Post argues that "once in the group, though, the power of group dynamics is immense, continually confirming the power of the group's organizing ideology and reinforcing the member's dedication to the cause." See Post, "Deterrence in an Age of Asymmetric Rivals," 171.

41. Lichbach, *Rebel's Dilemma*, 7.

42. Sewall, "Introduction to the University of Chicago Press Edition," xxv.

43. A good account can be found in Ricks, *Gamble*, 61–72. For an analysis see Koloski and Kolasheski, "Thickening the Lines."

44. Lebow and Stein, "Beyond Deterrence," 40.

45. There are generally three such bases: consequentialism, authority, and principles. Consequentialism is the easiest basis on which to operate, as the adversary is considering outcomes, and these are all potential as opposed to actual. They can therefore be altered by shifting the type of problem to be solved and hence the solution set from which it is appropriate to consider options. Authority-based frames are more difficult to affect because they are given to the adversary by another party, one to which the adversary has previously ceded decision authority. Thus that entity is a better target to attempt to influence. But another approach is to provide the adversary with a different authority than that to which he has ceded the decision—one that conflicts with the current authority frame—and make the case that he should switch. Finally, frames based upon principle are likewise difficult to affect, although the mechanism is similar: find a principle to which the adversary adheres and make the case that it applies more than the principle that provided the basis for defiance.

46. Jones and Libicki find that only 7 percent of terrorist groups that "ended" in their sample of 268 such groups were defeated by military force. "Militaries tended to be most effective when used against terrorist groups engaged in an insurgency in which the groups were large, well armed, and well organized. Insurgent groups have been among the most capable and

42

lethal terrorist groups, and military force has usually been a necessary component in such cases. Against most terrorist groups, however, military force is usually too blunt an instrument." See Jones and Libicki, *How Terrorist Groups End*, xiii–xiv.

47. In "Targeting the Leadership of Terrorist and Insurgent Movements," Langdon, Sarapu, and Wells found that killing the leader of a nonstate movement led to the disbanding or moderation of the movement in 61 percent of the cases that they examined (66, table 2).

48. Indeed, as Bernard Brodie famously put it, "Thus far the chief purpose of our military establishment has been to win wars. From now on its chief purpose must be to avert them. It can have almost no other useful purpose." See Brodie, "Implications for Military Policy," 76.

Bibliography

Abrahms, Max. "What Terrorists Really Want: Terrorist Motives and Counterterrorism Strategy." *International Security* 32, no. 4 (Spring 2008): 78–105.

Alexander, Gerard. "International Relations Theory Meets World Politics: The Neoconservative vs. Realism Debate." In *Understanding the Bush Doctrine: Psychology and Strategy in an Age of Terrorism*, edited by Stanley A. Renshon and Peter Suedfeld. New York: Routledge, 2007.

Allison, Graham T. *Essence of Decision: Explaining the Cuban Missile Crisis*. Boston: Little, Brown and Company, 1971.

Argo, Nichole. "The Banality of Evil: Understanding Today's Human Bombs." Unpublished policy paper, Preventive Defense Project. Stanford University, 2003.

Bloom, Mia M. "Palestinian Suicide Bombing, Public Support, Market Share, and Outbidding." *Political Science Quarterly* 119, no. 1 (2004): 61–88.

Brodie, Bernard. "Implications for Military Policy." In *The Absolute Weapon: Atomic Power and World Order*, edited by Bernard Brodie. New York: Harcourt, Brace and Company, 1946.

Brulé, David J. "Congress, Presidential Approval, and U.S. Dispute Initiation." *Foreign Policy Analysis* 4, no. 4 (October 2008): 349–70.

Bueno de Mesquita, Bruce, Alastair Smith, Randolph M. Siverson, and James D. Morrow. *The Logic of Political Survival*. Cambridge, MA: MIT Press, 2003.

Bush, Pres. George W. Address to a joint session of Congress, 20 September 2001. http://archives.cnn.com/2001/US/09/20/gen.bush.transcript/ (accessed 25 June 2009).

———. Graduation speech. US Military Academy, West Point, NY, 1 June 2002. http://georgewbush-whitehouse.archives.gov/news/releases/2002/06/20020601-3.html (accessed 3 June 2009).

Byman, Daniel. "The Decision to Begin Talks with Terrorists: Lessons for Policymakers." *Studies in Conflict & Terrorism* 29, no. 6 (June 2006): 403–14.

———. "Do Targeted Killings Work?" *Foreign Affairs* 85, no. 2 (March/April 2006): 95–112.

Crenshaw, Martha. "The Logic of Terrorism: Behavior as a Product of Strategic Choice." In *Origins of Terrorism: Psychologies, Ideologies, Theologies, States of Mind*, edited by Walter Reich. New York: Cambridge University Press, 1990.

Davis, Paul K., and Brian Michael Jenkins. *Deterrence and Influence in Counterterrorism: A Component in the War on al Qaeda.* Santa Monica, CA: RAND, 2002.

Faria, João Ricardo, and Daniel G. Arce M. "Terror Support and Recruitment." *Defence and Peace Economics* 16, no. 4 (August 2005): 263–73.

Freedman, Lawrence. "Strategic Coercion." In *Strategic Coercion: Concepts and Cases*, edited by Lawrence Freedman. New York: Oxford University Press, 1998.

Hanson, Victor Davis. "Islamicists Hate Us for Who We Are, Not What We Do." *Jewish World Review*, 13 January 2005. http://www.jewishworldreview.com/0105/hanson2005_01_13.php3 (accessed 24 May 2009).

Huth, Paul, and Bruce M. Russett. "Testing Deterrence Theory: Rigor Makes a Difference." *World Politics* 42, no. 4 (July 1990): 466–501.

Joint Forces Command. *Strategic Communication Joint Integrating Concept.* Version 0.5, 25 April 2008.

Jones, Seth G., and Martin C. Libicki. *How Terrorist Groups End: Lessons for Countering al Qa'ida.* Santa Monica, CA: RAND, 2008.

Kalman, Matthew. "Israeli Fence Puts 'Cage' on Villagers: More Palestinians Scrambling to Keep Barrier from Going Up." *San Francisco Chronicle*, 9 March 2004. http://www.sfgate.com

/cgi-bin/article.cgi?file=/chronicle/archive/2004/03/09/MNGIP5H0IL1.DTL (accessed 1 June 2009).

Koloski, Andrew W., and John S. Kolasheski. "Thickening the Lines: Sons of Iraq, a Combat Multiplier." *Military Review* 89, no. 1 (January–February 2009): 41–53.

Kydd, Andrew, and Barbara F. Walter. "Sabotaging the Peace: The Politics of Extremist Violence." *International Organization* 56, no. 2 (Spring 2002): 263–96.

Langdon, Lisa, Alexander J. Sarapu, and Matthew Wells. "Targeting the Leadership of Terrorist and Insurgent Movements: Historical Lessons for Contemporary Policy Makers." *Journal of Public and International Affairs* 15 (Spring 2004): 59–78.

Lebow, Richard Ned. *Between Peace and War: The Nature of International Crisis.* Baltimore, MD: The Johns Hopkins University Press, 1981.

Lebow, Richard Ned, and Janice Gross Stein. "Beyond Deterrence." *Journal of Social Issues* 43, no. 4 (Winter 1987): 5–71.

———. "Deterrence: The Elusive Dependent Variable." *World Politics* 42, no. 3 (April 1990): 336–69.

———. "Rational Deterrence Theory: I Think, Therefore I Deter." *World Politics* 41, no. 2 (January 1989): 208–24.

Levy, Jack S. "The Diversionary Theory of War: A Critique." In *Handbook of War Studies*, edited by Manus I. Midlarsky, 259–87. Boston, MA: Unwin Hyman, 1989.

Lichbach, Mark Irving. *The Rebel's Dilemma.* Ann Arbor, MI: University of Michigan Press, 1998.

Morgan, Patrick. *Deterrence Now.* Cambridge Studies in International Relations. Cambridge, UK: Cambridge University Press, 16 June 2003.

The National Security Strategy of the United States of America. Washington, DC: White House, October 2002.

Neumann, Peter R. "Negotiating with Terrorists." *Foreign Affairs* 86, no. 1 (January–February 2007): 128–38.

Pape, Robert A. *Dying to Win: The Strategic Logic of Suicide Terrorism.* New York: Random House, 2005.

Peters, Ralph. "Thou Shalt Kill." *Harper's Magazine*, January 2005, 19.

Post, Jerrold M. "Deterrence in an Age of Asymmetric Rivals." In *Understanding the Bush Doctrine: Psychology and Strategy in an Age of Terrorism*, edited by Stanley A. Renshon and Peter Suedfeld. New York: Routledge, 2007.

Ricks, Thomas. *The Gamble: General David Petraeus and the American Military Adventure in Iraq, 2006–2008*. New York: Penguin Press, 2009.

Schaub, Gary, Jr. "Deterrence, Compellence, and Rational Decision Making." PhD diss., University of Pittsburgh, 2003.

Schelling, Thomas C. *The Strategy of Conflict*. Cambridge, MA: Harvard University Press, 1960.

———. "What Purpose Can 'International Terrorism' Serve?" In *Violence, Terrorism, and Justice*, edited by R. G. Frey and Christopher W. Morris. New York: Cambridge University Press, 1991.

Sewall, Sarah. "Introduction to the University of Chicago Press Edition: A Radical Field Manual." In *The US Army/Marine Corps Counterinsurgency Field Manual*. Chicago: University of Chicago Press, 2007.

Simon, Herbert A. *Administrative Behavior: A Study of Decision-Making Process in Administrative Organization*. 3rd and 4th eds. New York: Free Press, 1976 and 1997. Originally published 1945.

Tolbert, Julian. *Crony Attack: Strategic Attack's Silver Bullet?* Maxwell AFB, AL: School of Advanced Air and Space Studies, 2003.

US Strategic Command. *Deterrence Operations Joint Operating Concept (DO JOC)*. Version 2.0, December 2006. http://oai .dtic.mil/oai/oai?verb=getRecord&metadataPrefix=html&id entifier=ADA490279.

Wohlstetter, Roberta. "Cuba and Pearl Harbor: Hindsight and Foresight." *Foreign Affairs* 43, no. 4 (July 1965): 691–707.

Chapter 4

From Observation to Action

Redefining Winning and
Sovereignty for the Information Age

CAPT John W. Bodnar, USNR, Retired, Science Applications International Corporation (SAIC)

Abstract: The transition from the industrial age to the information age requires a new way of thinking about adversaries and the international system. Instead of a bipolar world where security is defined as a win or lose situation, a multipolar world comes in many shades of gray that allow for win-win outcomes between adversaries. Instead of using industrial-age Newtonian and Clausewitzian physical models that focus on capability, new models are required to incorporate intent. If we (1) apply OODA (observe, orient, decide, act) loops, (2) redefine winning, (3) redefine sovereignty, and (4) think in terms of Mahanian win-win policies, we can begin to build the information-age analytical models needed to understand intent.

The process of developing models for intelligence analysis usually follows the simple six-question formula we all learned in high school. *Who, what, when,* and *where* are the bases upon which we then deduce *how* and *why.* As a defense community we customarily focus on *how*—capability and opportunity—rather than *why*—intent. Accordingly, we should not be surprised when the all-important question *what indicates a state actor's real intent?* is difficult to assess when we leave the question *why* as a mere add-on to our collection, reporting, and analysis.[1]

47

I suggest that we need to start our thinking with *why* and then build analytical methods and tools that can answer the other five questions in support of deducing *why*. Scientists assume that entities do what they do because of what they are while engineers and biologists assume that entities are what they are because of what they do. On the one hand, physical systems (such as hurricanes or tsunamis) operate without intent. On the other hand, engineering systems (such as torpedoes or ballistic missiles) are built for a purpose, and biological systems always operate with intent. A biological system strives to answer (either explicitly as in human thinking or implicitly as in natural selection) the following questions: What do I need to do to survive? What do I need to do to win (e.g., collect energy to build, work, and reproduce)? The organisms, species, and organizations that do those things best survive, live long, and prosper; the ones that do not, go extinct.

Therefore, I start with a simple intent model: Col John Boyd's decision cycle or OODA (observe, orient, decide, act) loop.[2] This simple model of organismal and organizational action is built on the premise that observations of the environment are the inputs to drive actions and decisions are required so that those actions can achieve the intent—to survive and win (described in greater detail elsewhere).[3] Indeed, we need to reexamine the core "intent" of every biological entity by starting with the essential processes that all living systems have evolved to insure winning and survival. Before we can adequately understand any nation's or organization's intent and how that might become a threat, we need to ask, how does it define survival? and how does it define winning? We also need to reexamine our own definitions, especially in light of the huge differences in international politics inherent in the transition from an industrial-age bipolar world to a multipolar community in the information age.

From Winning Two-Player, Zero-Sum Games to Winning Multiplayer, Non-Zero-Sum Games

US foreign policy is "scientific" in that it is based on the rules of Newtonian science—as adapted to statecraft by Clausewitz.[4]

Those rules, which were instrumental for preeminence during the industrial age, were based on the assumption that the world is continuous (can be subdivided infinitely as real numbers) and single-valued (having only a single solution or eigenvalue for any function), and they directly led to modeling the world as a two-player, zero-sum game. Since the end of the Cold War, this model has become outdated and must be replaced by one that is a multiplayer, non-zero-sum game. Such a change in basic assumptions for modeling US interactions in a multiplayer world will ultimately allow the United States to think outside the old industrial-age box and instead to think in a larger, new box in which the primary goal will be a win-win policy.

US Policy in Two-Player, Zero-Sum Games

The two-player, zero-sum game has been the basis for Western military-political thinking since Clausewitz. Clausewitz found that, just as in chess or checkers, there is a single winning strategy in a zero-sum game—the total annihilation of the other player. This strategy was first utilized by Generals Grant and Sherman against Confederate forces during the Civil War and has been successful in other cases where the United States could approximate the world situation as a two-player, zero-sum conflict: Allied versus Axis powers in World Wars I and II and the West versus the Soviet Union in the Cold War. In all these cases, the United States could easily think inside the box (fig. 4.1A), where both players—all disciples of Clausewitz—assumed a symmetrical win-lose and lose-win game.

As in chess or checkers, the only way to escape total annihilation for the loser in a zero-sum game is through unconditional surrender. Therefore, since Clausewitz, Western history has followed a pattern of matched zero-sum OODA loops where:

- Nations disagree.
- Nations solve their disagreement by armed conflict.
- The fight continues until one side surrenders unconditionally (or the cost of total war leads to an armistice that merely postpones the unconditional fight).
- The victor imposes conditions and assimilates the loser (or sets up a new government in the loser's nation).

A	US	THEM	Industrial Age World
US Foreign Policy	WIN	LOSE	
Zero-Sum Game	LOSE	WIN	Expected Clausewitzian Counter-Policy

B	US	THEM	The World Today
	WIN	WIN	
US Foreign Policy	WIN	LOSE	
Zero-Sum Game	LOSE	WIN	Expected "Axis of Evil" Counter-Policy
Non-Zero-Sum Game	LOSE	LOSE	Actual "Axis of Evil" Counter-Policy

Figure 4.1. Current US foreign policy is based on an industrial-age, Clause-witzian, two-player, zero-sum game model (A), which is no longer viable in an information-age, multiplayer, non-zero-sum environment (B).

The United States has been the most successful player in the two-player, zero-sum industrial-age world based on the symmetrical Clausewitzian strategy.

Foreign policy is relatively simple in a world where the two predominant players are evenly matched and think alike. To prevail in such an environment, one strives to have the "biggest stick" because the only endgame in such a world is when the most powerful military force imposes itself on all the other players, who are then assimilated into its sphere of influence and political dominance. This policy is the underlying assumption of current US foreign policy—an extension of "manifest destiny" (which has sometimes been called "white man's burden"—a patronizing view that Westerners must raise up their supposed non-Western inferiors). However, in the post–Cold War world, this kind of strategy is beginning to unravel because strategies in a multiplayer environment become asymmetric.

US Failures in Multiplayer, Non-Zero-Sum Games

US policy is at loose ends because it continues to play inside the box of the two-player, zero-sum game while the rest of the world is playing outside the box using multiplayer, non-zero-sum strategies. In this world, the United States cannot hope to find a stable endgame because its definitions of winning and losing are different from those of the other players and because it has continually ignored the additional strategies available in the non-zero-sum game (fig. 4.1B). Therefore, the United States is continually frustrated because the actions it takes are countered with unexpected out-of-the-box reactions that appear totally illogical within the framework of US industrial-age thinking. In this environment, small nations can exploit the dichotomy between two contradictory aspects of US foreign policy: US isolationism, in which the United States turns its back on any outside entanglements unless threatened, and US world involvement, in which the United States applies the Clausewitzian policy.

A successful strategy based on "playing chicken" with the United States outside the zero-sum-game box was first employed by Fidel Castro and Communist Cuba starting with the Cuban missile crisis. The strategy depends on the target state defining winning as survival and realizing the way to insure regime survival is a policy of "Yankee, go home!"—go away and leave us alone to determine our own policy inside our own sovereign borders. The United States has expected that starting diplomatic overtures based on regime change will cause the "underdog" to back down or surrender rather than fight, not realizing that the underdog, like every other biological entity, wants to win but wants more to survive. This initial overture calling for regime change to open diplomatic relations is like starting a meeting by saying, "Good morning, I've come today to plan your funeral." The Cuban underdog-versus-overdog rationale is a mismatch of interacting zero-sum and non-zero-sum OODA loops. This scenario, which subsequently has been used successfully by Vietnam, Somalia, and now North Korea, goes as described below.

The United States has a difference of policy with the target state and calls for regime change. The target state responds by doing nothing—staying within its own borders and not threatening anyone—and indicates that it will not take any action unless attacked. This is an attempt to prevail on US isolationism to craft a win-win scenario, that is, the target nation wins by being free of US influence while the United States wins by not being threatened.

51

The United States then reacts in its world involvement mode and misinterprets the target state's response as a lose-win strategy. The United States reiterates its win-lose policy and indicates that nonsurrender by the target state will only lead to a US military (or economic) victory and regime change—total annihilation of the target state's current government.

The target state then amplifies that it will continue to do nothing outside its borders and will not threaten anyone—unless attacked. This changes the target nation's policy to an apparent lose-lose scenario because it courts US military action.

The United States misinterprets the new lose-lose scenario as totally irrational because such a strategy is "not allowed" within the zero-sum box. It interprets the new strategy as a lose-win counterpolicy but cannot understand how the target nation could be so stupid as to believe it could defeat the United States.

At this point, the United States has three options in its zero-sum thinking:

1. *Do nothing.* This is a win-win scenario under a US foreign policy of isolationism. However, this becomes a lose-win scenario under a US policy of world involvement because failure to dominate the other player in a zero-sum game is a "lose" by Clausewitz's definition. In either case, it is a "win" for the target nation.

2. *Use low-level military force* against the target nation and hope that will cause the target nation regime to fail. The Bay of Pigs invasion, initial actions in Vietnam, and deployment of troops in Somalia were examples of this strategy.

3. *Use massive military force* against the target nation to topple its regime. The escalated Vietnam War, Desert Storm, and the initial strategy in Afghanistan and Iraq fit this category.

The target nation will do everything it can to divert the conflict back to its perceived win-win strategy—"if you leave me alone, I'll leave you alone." For this reason scenario one above is always acceptable to the target nation—but rarely to the United States.

For scenarios two and three, the target nation still can win if it can make the cost of those scenarios high enough that US isolationism trumps world involvement and the United States goes home—or at least leaves the current regime in power. The United States continually misinterprets the target nation's actions at this step since the United States invariably interprets a threat of military action by the target nation as a doomed attempt at a lose-win strategy rather than understanding that causing the United States to go away without toppling the target nation regime—no matter the cost—is considered a win by the target nation.

Cuba was the first to win against the United States in the non-zero-sum game at the end of the industrial age. And several subsequent showdowns have followed the same pattern (for example, Vietnam, Beirut Marine barracks attack, Somalia). By thinking in a win-lose and lose-win box, the United States has repeatedly failed to recognize that an adversary's best chance against the United States in any conflict is to go for a lose-lose strategy, which can be converted into a win-win outcome *if* that nation can force the United States out of its world involvement mode and back into its isolationist mode. The most frustrating part of this strategy is that US administrations find themselves faced with the possibility that they can win only by losing—which is unacceptable within the zero-sum game.

Clearly, new ways of thinking, which require a total re-examination of US strategy, are needed for the information age. What constitutes a win and a loss in military-political policy? How can we balance the dichotomy of the traditional US policy of isolationism and the new policy of world involvement? Will the US world involvement policy remain a variation on manifest destiny and white man's burden, or can the United States come up with a new international policy that is not seen by the rest of the world as a revival or extension of these policies?

From Sovereignty in a Bipolar World to Sovereignty in a Multinational Community

Inherent in any society or community is the dichotomy between individual freedom and the common good. That dichotomy can be addressed in many ways, but the two extremes are absolute individual freedom (which equals anarchy) and absolute order (which equals totalitarianism). The US Constitution recognizes that there must be a dynamic balance of both and provides mechanisms to insure the best balance. As we enter the information age, the world is forming a global community whose challenge is to address the question of freedom versus order in a society of individual nations the way the US Constitution addresses freedom versus order for individual citizens, communities, and states. In the past, national sovereignty has implied total freedom of action completely independent of the actions of other nations and total freedom from outside interference into affairs within sovereign borders. Therefore, as the world moves toward an interdependent global community, national policy must begin to reflect the information revolution: one nation's actions are not totally independent of other nations, and the concept of national sovereignty is changing dramatically.

Balancing Freedom and Order

The United States has always acted as an individual nation, but as a world community evolves, the United States must recognize the dichotomies inherent in being a member of any community. With this in mind, I present a new set of problems the United States is facing for the first time in history, caused by the emerging global community. National policies based on traditional definitions of national sovereignty, which worked very well in the past, could lead to disaster in the next few decades. Every society or community must make choices on a number of basic issues between the two extremes of total freedom and total order.

TOTAL FREEDOM	vs.	CHECKS AND BALANCES	vs.	TOTAL ORDER
Instability	vs.	Dynamic Stability	vs.	Stability
Anarchy	vs.	Constitutional Law	vs.	Totalitarianism
Utopia	vs.	US, United Kingdom, France	vs.	Nazi Germany, USSR
al-Qaeda, initial efforts in Iraq and Afghanistan		vs.		Iraq under Saddam, North Korea
Independence		vs.		Interdependence
Rights (owned by individual)		vs.		Privileges (owned by state)
Information access		vs.		Information control

54

The strength of a constitutional government is in its balance between the two extremes: protecting individual rights to the maximum extent possible while still providing for the common good. There are many variations on the theme (for example, the United States, the nations of the European community, the nations of the former British Commonwealth, India, and Japan). The common principle is that any true constitutional government is based on a separation of powers and a tension between the individual and the state—leading to a dynamic stability in which order imposed by the constitution can evolve in a changing world through self-imposed mechanisms for amendment. Thus, in many ways, the stability inherent in the US Constitution is caused by the clauses within the Constitution that allow each generation to modify and reinterpret it but provide restraints on how much or how fast those changes can occur.

Redefining Intent the Mahanian Way

I suggest that a blueprint for the road ahead can be found in the teachings (fig. 4.2) of Alfred Thayer Mahan—a blueprint that encompasses definitions for both "winning" and "sovereignty" applicable for the multiplayer, non-zero-sum game in the information age (a world community) and that provides a basis for understanding intent for both "us" and "them."[5]

	US	THEM	Information-Age World
US Foreign Policy?	WIN	WIN	Expected Mahanian Counter-Policy
	WIN	LOSE	
	LOSE	WIN	
Non-Zero-Sum Game	LOSE	LOSE	

Figure 4.2. The teachings of Alfred Thayer Mahan can provide a strategy to overcome the mismatch between US foreign policy and the information-age global community.

Clausewitz assumes a world at war and, therefore, tells how to win wars. Mahan assumes that the job of the Navy is to keep commerce flowing. War is sometimes necessary to assure that happens, but the ultimate goal is "innocent passage" through all the sea-lanes across the world. By assuming that peace is impossible, disciples of Clausewitz do not even consider it an option. By assuming that conflict-free commerce is the goal, disciples of Mahan can maintain peace.

Clausewitz's book was on *war*:

- Inherently two-player, zero-sum game.

- I win–you lose or you win–I lose.

- Take out the primary objective, then move on to the next, considering each independently, one at a time—for example, the recent US hit list of Osama bin Laden, Afghanistan, and Iraq, with North Korea and Iran next on the list.

Mahan's book was on *sea power*:

- Inherently non-zero-sum game.

- If I can maintain my business—keeping the seas free for what I want to do—it does not really matter if that allows you to do the same.

- If I win, I do not care whether you win or lose. This inherently adds the win-win and lose-lose scenarios to the mix—for example, the British building the Commonwealth and building democracy in India.

The most telling of Mahan's sea stories is about why Hannibal crossed the Alps. Because the Roman navy was so strong, crossing the Straits of Messina to march directly on Rome was not an option, so Hannibal needed to march thousands of miles out of his way via the Alps to get to Rome. Thus the Roman navy was a huge factor in Hannibal's defeat—even though it never fought a battle. Mahan's genius was to follow that thread through to the maritime strategy which was the basis of *Pax Britannica* for several hundred years and which, through Mahan's teachings, has been inherited by the US Navy.[6] How can we get the United States to stop thinking like Clausewitz and start thinking like Mahan?

Conclusion

Making the best decisions in any OODA loop requires an understanding of intent, both one's own and that of the other players. What you do and what you *can* do are enabled or limited by your culture. You can predict what the opponent will do if you know his culture. Knowing that will help you either destroy him or join with him to build a better world. Whether you choose to destroy or build is dependent on your particular worldview and culture—and on core values such as how you define "winning" and "sovereignty" and what kind of "game" you think you are playing.

If your cultural model of interaction is a two-player, zero-sum game (like Clausewitz's), your strategy for any game is absolute, total annihilation of the opponent; if your model is a multi-player, non-zero-sum game (like Mahan's), your strategy is to find the right balance for your side to survive. Therefore, how you perceive the other players' strategies is colored by your own assumptions. I believe it is time for the United States to rethink and redefine its assumptions.

Notes

(All notes appear in shortened form. For full details, see the appropriate entry in the bibliography.)

1. Bodnar, *Warning Analysis for the Information Age*, 14; and Grabo, *Anticipating Surprise*, 17, 164.
2. Boyd, "Discourse on Winning and Losing."
3. Bodnar, *Warning Analysis for the Information Age*; Bodnar, "Information Age Decision-Making"; and Bodnar, "Making Sense of Massive Data by Hypothesis Testing."
4. Detailed in Bodnar, *Warning Analysis for the Information Age*.
5. Mahan, *From Sail to Steam*; and Mahan, *The Influence of Sea Power upon History*.
6. This strategy was first outlined in *The Influence of Sea Power upon History* and fleshed out through his later teachings.

Bibliography

Bodnar, John W. "Information Age Decision-Making: Reaffirming Marine Corps Leadership Traditions." *Marine Corps Gazette* 89, no. 4 (2005): 20–24.

———. "Making Sense of Massive Data by Hypothesis Testing." Presentation. 2005 International Conference on Intelligence Analysis, McLean, VA, 2–4 May 2005. https://analysis.mitre

.org/proceedings/Final_Papers_Files/124_Camera_Ready _Paper.pdf.

———. *Warning Analysis for the Information Age: Rethinking the Intelligence Process.* Washington, DC: Joint Military Intelligence College, 2003. http://www.dia.mil/college/pubs /pdf/3245.pdf.

Boyd, Col John R. "A Discourse on Winning and Losing." Collection of unpublished briefing slides, August 1987. Muir S. Fairchild Research Information Center, Maxwell AFB, Document No. M-U 43947.

Grabo, Cynthia M. *Anticipating Surprise: Analysis for Strategic Warning.* Washington, DC: Joint Military Intelligence College's Center for Strategic Intelligence Research, 2002.

Mahan, Alfred Thayer. *From Sail to Steam: Recollections of Naval Life.* New York: Harper & Brothers, 1907.

———. *The Influence of Sea Power upon History, 1660–1783.* 5th ed. Boston: Little, Brown and Co., 1894. Reprint, Mineola, NY: Dover Publications, 1987.

PART 2

Academic Perspective
Theory and Research in Gauging Intent

Chapter 5

Anthropological Reflections on Motive and Intent and the Strategic Multilayer Assessment Typology

Lawrence A. Kuznar, PhD, NSI, Inc.

Abstract: This paper provides a review of basic motivating factors recognized by anthropologists that help explain intent. Intentions are influenced by a variety of factors, from shared cultural values to evolutionary psychology. These factors are then related to a general sociocultural typology used in the Strategic Multilayer Assessment (SMA) effort to structure analyses of human, social, cultural, and behavioral factors. Various anthropological perspectives on the sources of motivating factors that influence intent are explored including structuralism, interpretivism/symbolic anthropology, postmodernism, culture and personality, human behavioral ecology, and discourse analysis. Brief examples of how these perspectives could be used to interpret the intentions of al-Qaeda are included. While the diversity of anthropological approaches illustrates the lack of a unified approach to understanding intent, each approach provides a window on how an actor's intent could be judged.

Intention implies a conscious, desired end state an individual or an organization may strive to achieve. Intentions are influenced by a variety of factors, from shared cultural values to evolutionary psychology. In this chapter, I review basic motivating factors considered by anthropologists and relate these to a general sociocultural typology[1] currently used in Strategic Multilayer Assessment (SMA) efforts to structure analyses of human, social, cultural, and behavioral factors.[2]

The chapter begins with an overview of the SMA sociocultural typology and continues with various anthropological perspectives on the sources of motivating factors that influence intent. This chapter cannot provide an exhaustive discussion of all relevant perspectives, nor can it fully describe the theory and method of each approach. However, it can provide an overview of several important anthropological approaches relevant for assessing the intentions of individuals and organizations. The primary approaches covered include structuralism, interpretivism/symbolic anthropology, postmodernism, culture and personality, human behavioral ecology, and discourse analysis. In each case, I provide very brief examples of how these perspectives could be used to interpret the intentions of al-Qaeda.

The SMA Sociocultural Typology

The SMA sociocultural typology was developed to provide a *generalizable* typology for military and intelligence analysts and planners to characterize sociocultural systems. This typology is based on comparative analysis of academic sociocultural typologies, sociocultural typologies produced for the US military and intelligence community, and standard approaches to examining levers of power (diplomatic, informational, military, and economic) and their effects (political, military, economic, social, informational, and infrastructural). The typology has academic roots in the work of A. R. Radcliffe-Brown, Bronislaw Malinowski, Julian Steward, Edward Hall, Marvin Harris, David Wilson, and the Human Relations Area Files Outline of Cultural Materials.[3]

The typology is organized in a hierarchy that includes increasingly detailed levels (fig. 5.1). The levels include five fundamental categories (interests, capabilities, context, decision-making psychology, and language) with 10 high-level variables: *interests*—(1) motivating factors, religion, and ideology, (2) social identity, (3) objectives; *capabilities*—(4) economy, technology, and other capabilities; *context*—(5) roles/life cycle, (6) demography, (7) political and social organization, (8) environmental and historical context and other actors; (9) *decision-making psychology*; and (10) *language*. Subcategories can be added as necessary. This typology is designed to provide a broad coverage of relevant sociocultural factors at sufficient detail to facilitate finer inquiries into the exact data required for a sociocultural analysis; it is not an exhaustive list of all possible sociocultural variables, and it is intended to be a "living document," revised as SMA efforts grow.

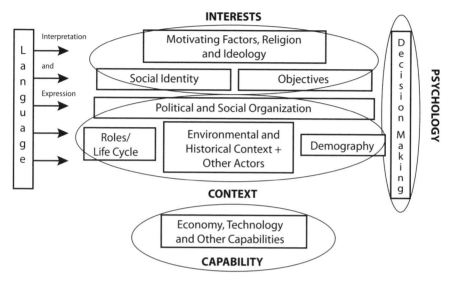

Figure 5.1. SMA sociocultural typology. (Created by author)

From an anthropological perspective, motivating factors, religion, ideology, contextual factors, and language are the primary cultural influences on an actor's intent and are therefore the focus of this chapter.

Motivating Factors, Religion, and Ideology

Many scholars assume that our intentions are derived from deeply held beliefs, often encoded in shared values and religious or other ideological belief systems. The SMA sociocultural typology captures many of these elements in the "motivating factors, religion, and ideology" variable. This common idea rests upon deep theoretical and philosophical assumptions about ideology and human behavior that are not resolved in the anthropological community. Major influences on the "ideology to intent and behavior" paradigm include philosophical structuralism, interpretivism/symbolic anthropology, and recent trends in postmodern thought. These approaches are not necessarily mutually exclusive and often bear similar analytical results.

Structuralism, an approach developed in anthropology by Claude Lévi-Strauss, is based on the axiom that humans perceive the world in terms of binary opposites.[4] These opposites structure human experience, which in turn reproduces these cognitive structures. By analyzing myths, social structures, and behaviors, one can reveal a deep cultural grammar that constitutes the structured meanings of a culture. Furthermore, the practitioners of culture are unaware of these deep meanings. Since these meanings structure perception and experience, they therefore provide the values upon which intentions would be based. From a structuralist perspective, intent would have to be derived from deeply seated cognitive structures and values that may be revealed through the analysis of discourse and behavior. For example, a structuralist analysis of al-Qaeda's worldview would focus on its Manichean division of the world into good (an Islamic world according to al-Qaeda's dictates) versus evil (a corrupted West).

The interpretivist/symbolic school of anthropology focuses on how people construct meaning through the use of symbols and symbolic acts. Perhaps the most influential scholar in this field was Clifford Geertz, who advocated "thick description," or the teasing apart of cultural meanings through the interpretation of symbols and actions in a particular cultural context.[5] According to this perspective, meaning cannot be discovered outside of its particular cultural context. The same symbols and actions in another context could take on very different meanings. Both interpretivist and structuralist approaches rely on the interpretation of the meanings of events, symbols, and actions, but interpretivist approaches privilege the context-specific construction of meaning by actors and are less (or not at all) concerned with uncovering deep, transcendent systems of meaning. Intent from this perspective would be judged from the context of cultural meanings surrounding an event. As an example, a thick description of the 1993 Twin Towers bombing might include the following themes: Islam versus the West, Third World versus globalizing financial institutions, and intent to increase the intensity of attacks against symbolic US targets.

Structuralist and interpretivist schools heavily influenced postmodern anthropology of the late twentieth and early twenty-first century. Postmodernism, sometimes called poststructuralism, is in part a rejection of structuralist attempts to uncover deep meanings based on fundamentals in human perception. Postmodernists stress the idiosyncratic and context-dependent nature of culture, as well as the active manipulation of culture and symbols by individuals for political ends.[6] Postmodernists argue that culture and meaning are so highly context dependent, and the ulterior motives of actors so permeate human discourse, that no event, action, or utterance can be innocently interpreted; therefore, these interpretations cannot provide the basis for generalizations about culture. Furthermore, in deconstructing a text to uncover its meaning, one uncovers so many layers of meaning, and one's own interpretations are so dependent on one's own biases, that no stable meaning is to be discovered.[7] Despite the seemingly nihilistic logic of postmodern analysis, it has become the major interpretive framework used by anthropologists to discuss the meanings of people's actions and analyze their intent. A postmodern analysis of Osama bin Laden's 1998 declaration of war against the Americans and "crusaders" would highlight resistance against Western domination, the suffering of Iraqis due to Western-imposed sanctions, and the intention to attack US targets.

Cultural Context and Intent

Postmodern and interpretivist schools heavily stress cultural context. Other anthropological approaches also derive their analysis of intent and meaning from context but are based on more specific aspects of cultural context. Examples include the "culture and personality" school of thought and human behavioral ecology.

The basic argument of the culture and personality school is that child-rearing practices, which vary across cultures, lead to differing psychological development processes that result in different modal personality types among cultures.[8] The SMA sociocultural typology captures these dimensions in the roles/life-cycle variable. A classic example of a culture and personality analysis is Ruth Benedict's study of Japanese culture, commissioned by the Office of War Information during World War II.[9] Culture and personality studies, however, have been difficult to test scientifically, leading to controversies in how they are used. More recent research on the effects of child rearing on culture is more scientifically based and nuanced, but generalizations are still made with great caution.[10] Intent from this perspective would be identified by observing how adult behavior is modeled to children and how people resolve challenges in the course of their psychological development. A culture and personality analysis of bin Laden's intentions would include analysis of the difficulties he may have encountered as one of many sons where expectations were very high, the separation of the sexes among Arabic adolescents, and the importance of being socialized with shame and honor values in intensifying his grievance against the US and Saudi governments.

Human behavioral ecology approaches are based in Darwinian evolutionary theory and posit a deep underlying human motivation to acquire the resources needed to reproduce (territory, food, shelter, mates, and security).[11] Many of the variables relevant to these motivations are captured in the SMA demography variable. Cultural success theory is a variant of human behavioral ecology that notes how reproductive benefits are often the indirect result of men attaining high social status through success in competitive ventures such as raiding.[12] Most human behavioral ecology approaches are grounded in models of individual, selfishly motivated behavior. Intent is understood as the intention to acquire resources and mates and to favor closely related kin because they share one's genes. Classic human behavioral ecology analyses of intent would suggest the recruiting power terrorist organizations have among young men with few ways to gain the status and resources necessary to marry in traditional Arabic society.[13]

An alternative human behavioral ecology approach assumes instead that humans are inherently social and therefore tend to act altruistically toward members of their group (family, lineage, tribe, ethnic group, religious group, etc.).[14] For instance, Peter Richerson and Robert Boyd demonstrated a correlation between the combat effectiveness of units in World War II and how closely unit recruitment and training approximated kin relations.[15]

Intent from this perspective is driven by a desire to help and defend members of one's own group. Analyses of bin Laden's 1998 declaration might indicate concern for fellow Muslims as a motivating factor that led to his intent to continue attacking US targets. Studies of suicide bombers that emphasize bombers' altruistic motives fall within this mode of explanation.[16]

Language and Intent

Varieties of linguistic discourse analysis have also been influential in anthropological attempts to analyze intent. These analyses relate to the language category of the SMA typology and recognize that language influences all aspects of culture. The simplest forms of discourse analysis are seemingly straightforward interpretations of what people say or what their doctrines state; sometimes people telegraph their intentions quite transparently. For instance, in a 1998 interview, referring to the United States, Osama bin Laden stated that "it is the duty of every Muslim to struggle for its annihilation." One could interpret this as a straightforward statement of intent to kill all Americans. However, what people say and what they do may not be the same, and people address different audiences differently, complicating the straightforward interpretation of discourse. More sophisticated approaches to discourse analysis are exemplified by the work of George Lakoff and Teun van Dijk.

Lakoff argues that the meaning of discourse is derived from overlapping metaphors.[17] According to Lakoff, a concept rarely has a clear-cut declarative meaning but rather derives its meaning in association with other concepts, which often serve as metaphors. Van Dijk's critical discourse analysis similarly seeks to uncover meaning from metaphorical association of concepts. Much of van Dijk's research concerns the euphemization (self-praise) of one's own group (in-group) and the derogation of other groups (out-groups) and therefore directly relates to studies of conflict. Van Dijk has identified 27 rhetorical devices used to express inequality and moral distinctions among groups.[18] Intent from a discourse analysis perspective would be derived from analysis of the metaphorical associations of concepts in an actor's discourse. And so, in bin Laden's 1998 declaration, derogative terms such as "devil's army" conjure demonic metaphors for the US military, and the term "crusader" links Western powers with rapacious aggression in the minds of Arabic readers.

Summary

This brief overview of anthropological perspectives on intent demonstrates the diversity of anthropological approaches, differences in theoretical underpinnings, and a consequent lack of a unified approach. However, each approach provides a window on how an actor's intent could be judged. Some approaches are clear as to their data requirements and methodology (human behavioral ecology and linguistic discourse analysis), others are less clear (culture and personality, structuralism, and interpretivism), and yet others eschew method altogether (postmodernism). The approaches relate to different aspects of a general sociocultural typology such as that used by SMA. Probably the greatest advantage of an anthropological approach is in its increased awareness of the great variety of sources for intent (ideology, metaphoric associations, group values, individual motives, and evolutionary drives) and the types of sociocultural data required to gauge intent.

Notes

(All notes appear in shortened form. For full details, see the appropriate entry in the bibliography.)

1. Chesser, *Deterrence in the 21st Century.*
2. SMAs are conducted in the Rapid Response Transition Office of the Directorate of Defense Research and Engineering in the Office of the Secretary of Defense.
3. Radcliffe-Brown, *Structure and Function in Primitive Society*; Malinowski, *Scientific Theory of Culture and Other Essays*; Steward, *Theory of Culture Change*; Hall, *Silent Language*; Harris, *Cultural Materialism*; Wilson, *Indigenous South Americans of the Past and Present*; and Murdock, *Outline of Cultural Materials.*
4. Lévi-Strauss, *Structural Anthropology.*
5. Geertz, *Interpretation of Cultures.*
6. Clifford, "On Ethnographic Authority"; Herzfeld, *Anthropology*; Rabinow, "Representations Are Social Facts"; Rosaldo, *Culture and Truth*; and Tyler, "A Post-Modern In-Stance."
7. See especially Tyler, "A Post-Modern In-Stance."

8. Benedict, *Patterns of Culture*; Mead, *Coming of Age in Samoa*; and Mead, *Sex and Temperament in Three Primitive Societies*.

9. Benedict, *Chrysanthemum and the Sword*.

10. Gregg, *Middle East*.

11. Chagnon, "Reproductive and Somatic Conflicts of Interest"; and Gat, "Human Motivational Complex," Parts I and II.

12. Chagnon, "Life Histories, Blood Revenge, and Warfare"; and Irons, "Cultural and Biological Success."

13. Simons, "Making Enemies."

14. Richerson and Boyd, *Not by Genes Alone*.

15. Richerson and Boyd, "Complex Societies."

16. Pedahzur, Perliger, and Weinberg, "Altruism and Fatalism."

17. Lakoff and Johnson, *Metaphors We Live By*.

18. van Dijk, "Politics, Ideology and Discourse."

Bibliography

Benedict, Ruth. *The Chrysanthemum and the Sword: Patterns of Japanese Culture*. New York: Houghton Mifflin, 1946.

——. *Patterns of Culture*. Boston: Houghton Mifflin, 1934.

Chagnon, Napoleon A. "Life Histories, Blood Revenge, and Warfare in a Tribal Population." *Science* 239, no. 4843 (26 February 1988): 985–92.

——. "Reproductive and Somatic Conflicts of Interest in the Genesis of Violence and Warfare among Tribesmen." In *The Anthropology of War*, edited by Jonathan Haas, 77–104. Cambridge: Cambridge University Press, 1990.

Chesser, Nancy, ed. *Deterrence in the 21st Century: An Effects-Based Approach in an Interconnected World*. Report prepared for US Strategic Command Global Innovation and Strategy Center (USSTRATCOM/GISC), Strategic Multilayer Assessment Team, 2007.

Clifford, James. "On Ethnographic Authority." *Representations* 2 (Spring 1983): 118–46.

Gat, Azar. "The Human Motivational Complex: Evolutionary Theory and the Causes of Hunter-Gatherer Fighting, Part I. Primary Somatic and Reproductive Causes." *Anthropological Quarterly* 73, no. 2 (April 2000): 20–34.

——. "The Human Motivational Complex: Evolutionary Theory and the Causes of Hunter-Gatherer Fighting, Part II. Proximate, Subordinate, and Derivative Causes." *Anthropological Quarterly* 73, no. 2 (April 2000): 74–88.

Geertz, Clifford. *The Interpretation of Cultures.* New York: Basic Books, 1973.

Gregg, Gary S. *The Middle East: A Cultural Psychology.* Oxford: Oxford University Press, 2005.

Hall, Edward Twitchell. *The Silent Language.* 1959. Reprint, Greenwich, CT: Fawcett Publications, 1966.

Harris, Marvin. *Cultural Materialism: The Struggle for a Science of Culture.* New York: Vintage Books, 1979.

Herzfeld, Michael. *Anthropology: Theoretical Practice in Culture and Society.* Oxford: Blackwell Publishing, 2001.

Irons, William. "Cultural and Biological Success." In *Evolutionary Biology and Human Social Behavior,* edited by Napoleon A. Chagnon and William Irons, 257–72. North Scituate, MA: Duxbury Press, 1979.

Lakoff, George, and Mark Johnson. *Metaphors We Live By.* Chicago: University of Chicago Press, 1980.

Levi-Strauss, Claude. *Structural Anthropology.* Garden City, NY: Doubleday, 1963.

Malinowski, Bronislaw. *A Scientific Theory of Culture and Other Essays.* Chapel Hill, NC: University of North Carolina Press, 1944.

Mead, Margaret. *Coming of Age in Samoa: A Psychological Study of Primitive Youth for Western Civilization.* New York: Blue Ribbon Books, 1928.

———. *Sex and Temperament in Three Primitive Societies.* New York: W. Morrow, 1935.

Murdock, George P. *Outline of Cultural Materials.* 5th ed. with modifications. New Haven, CT: Human Relations Area Files, 2004. http://www.yale.edu/hraf/Ocm_xml/newOcm.xml.

Pedahzur, Ami, Arie Perliger, and Leonard Weinberg. "Altruism and Fatalism: The Characteristics of Palestinian Suicide Terrorists." *Deviant Behavior* 24 (2003): 405–23.

Rabinow, Paul. "Representations Are Social Facts: Modernity and Post-Modernity in Anthropology." In *Writing Culture: The Poetics and Politics of Ethnography,* edited by James Clifford and George E. Marcus, 234–61. Berkeley: University of California Press, 1986.

Radcliffe-Brown, A. R. *Structure and Function in Primitive Society.* New York: Free Press, 1965.

Richerson, Peter J., and Robert Boyd. "Complex Societies: The Evolutionary Origins of a Crude Superorganism." *Human Nature* 10, no. 3 (1999): 253–89.

———. *Not by Genes Alone: How Culture Transformed Human Evolution.* Chicago: University of Chicago Press, 2005.

Rosaldo, Renato. *Culture and Truth: The Remaking of Social Analysis.* Boston: Beacon Press, 1989.

Simons, Anna. "Making Enemies: An Anthropology of Islamist Terror, Part I." *American Interest* 1, no. 4 (Summer 2006): 6–18.

Steward, Julian. *Theory of Culture Change: The Methodology of Multilinear Evolution.* Urbana, IL: University of Illinois Press, 1955.

Tyler, Stephen A. "A Post-Modern In-Stance." In *Constructing Knowledge: Authority and Critique in Social Science*, edited by Lorraine Nencel and Peter Pels, 78–94. London: Sage, 1991.

van Dijk, Teun A. "Politics, Ideology and Discourse." In *Encyclopedia of Language and Linguistics*, edited by Ruth Wodak, "Language and Politics" section. New York: Elsevier, 2004.

Wilson, David J. *Indigenous South Americans of the Past and Present: An Ecological Perspective.* Boulder, CO: Westview Press, 1999.

Chapter 6

Psychology

Policy Makers and Their Interpretations Matter

Margaret G. Hermann, PhD, Moynihan Institute of Global Affairs, Syracuse University

Abstract: Learning how policy makers view what is happening to them is critical to understanding how governments are likely to act. This statement is particularly true for those policy makers who have the authority to commit the resources of the government. Such policy makers' interpretations appear to be influenced by their beliefs, their experience, their thoughts about the chances of losing or winning in a situation, and their view of the constraints under which they operate. In effect, their rationality is bounded or conditioned by such characteristics. Depending on the nature of these characteristics, policy makers are likely to deal with situations on a case-by-case basis or to be directly and immediately influenced by their own personal predispositions. Researchers are developing techniques to assess these characteristics by taking advantage of the growing number of Web-based speeches, interviews, and writings available on leaders from around the world. Some of these techniques have been turned into software programs that enable the analyst to examine all available materials and contextualize the resulting profiles.

People Matter

Richard Snyder and his colleagues argued in an influential monograph that people matter in international affairs and launched the study of foreign-policy decision making.[1] Indeed, they contended that policy makers who perceive and interpret events and whose preferences become aggregated in the decision-making process shape what governments and institutions do in the foreign policy arena. People affect the way foreign policy problems are framed, the options that are considered, the choices that are made, and how policy gets implemented. To bolster their claims, Snyder and his associates brought research from cognitive, social, and organizational psychology to the attention of scholars interested in world politics.

Why are people important? For one thing, foreign policy problems are generally complex and ill structured. They demand interpretation for several reasons: there is no "correct" answer, there may be uncertainty about the nature and salience of the problem, what is happening may need to be placed in some structure or frame, and there may be value trade-offs.[2] How decision makers define and represent the problem may or may not match how an outside observer views it. In fact, research has shown that around 75 percent of the time policy makers involved with foreign policy issues disagree about the nature of the problem, the options that are feasible, or what should happen.[3] Note how the same event—the terrorist attacks of 11 September 2001—was framed differently by leaders in Britain and the United States. Prime Minister Tony Blair announced at the Labour Party conference just hours after the Twin Towers collapsed that we had just experienced a crime against civilization—police, courts, and justice were the instruments for dealing with what had happened; Pres. George W. Bush framed the event as an attack on America and pronounced a war on terror, engaging the military and calling forth nationalism.

Moreover, the people in governments change, and with each change can come a difference in perspective. Consider that in the past decade (1998–2008), the 29 Asian countries bordering the Pacific Rim have had 133 governments with 124 different leaders.[4] Often these governments are coalitions of parties. Some 60 percent of the leaders during this decade who came to power with an agenda and intention to control the policy-making process by defining what was important were overturned through irregular means—by votes of no confidence, calls for early elections, parties withdrawing from a coalition, and coups.

Interestingly, only 21 percent of leaders who believed in the use of informal power and preferred to work behind the scenes to make policy experienced an irregular regime change—when regime change occurred, it was regular and planned. Leadership style and strategy affected longevity in office as well as how influential the leaders' views were in what was considered a problem and who was involved in making policy.

Research has also shown that there is a contraction of authority to those most accountable for policy in crisis situations.[5] Such a contraction appears to happen regardless of the type of political system.[6] People and their interpretations of what is happening become more important in crisis situations, which are usually considered to involve a serious threat to the values and interests of the government, allow little time for decision making, and come as a surprise. Indeed, in a study of 81 international crises, identified as such by journalists, historians, and political scientists, how the policy makers viewed the amount of time available to them and the degree of surprise in the situation led to different decision-making processes.[7] When they viewed themselves as having little time and were surprised (a 9/11-type of event), the policy makers pushed to frame the event quickly, reach a rapid consensus on what to do, and then implement their decision with little interest in, or reaction to, feedback regarding what they were doing—either positive or negative. They engaged in path-dependent behavior. However, if policy makers thought themselves to have a little more time in which to decide, even if they were caught by surprise (the US reaction to the Iraqi invasion of Kuwait, for example), they became more innovative—searching for information and expertise that could help them ascertain what was happening and trying to think outside the box. The data suggest that the public is much more pleased with the first set of policy makers than the second since the first appears decisive while the second takes time to discuss and consider the situation out of the public's eye, seeming not to be taking any action even though in the long run their actions are generally more effective.

Rationality Is Bounded

Nobel laureate in economics Herbert Simon pushed us further by arguing that not only do people matter, but they do not necessarily act rationally.[8] His experimental studies of decision making indicated that rationality is bounded by how the people involved process information, what they want, the ways in which they represent the problem, their experiences, and their beliefs. In effect, decision makers "do not have unlimited time, resources, and information" to make choices that maximize their movement toward their goals.[9] They "satisfice"—settling for the first acceptable option rather than pushing for ever more information and a more optimal choice. "People are, at best, rational in terms of what they are aware of, [but] they can be aware of only tiny, disjointed facets of reality."[10] It becomes important to learn about the foreign policy makers' "view of reality" as their preferences, so defined, will shape their actions. Building on this idea, for example, we have ascertained that (1) beliefs are like possessions and are relinquished only reluctantly, (2) decision makers' interpretations of the nature of a situation appear to determine how risk prone or risk averse their actions are likely to be, (3) prior experience and knowledge about a problem have been found to shape cognition and focus decision making while a lack of such expertise leaves decision makers to rely on their personal predispositions, and (4) decision makers perceive and react to political constraints differently, depending on their leadership style.

In 1969 Alexander George proposed that policy makers are guided by an "operational code"—a set of philosophical and instrumental beliefs that set the parameters for their actions. These beliefs help to define what is viewed as a problem and what options are seen as viable within that particular orientation to politics. For instance, consider what might be the differences in the actions of policy makers who believe that conflict is endemic to politics and those that view it as generally temporary and the result of misunderstanding. For the first type, the world is full of threats, vigilance is necessary as control and predictability are limited, and all actors are potential rivals, whereas with a misunderstanding there is the opportunity to change the other's view and thus to control any escalation as well as to establish a climate that can foster negotiation.[11] The operational code has proven particularly useful in examining the strategic interaction among leaders of countries during a crisis.[12] By determining the leaders' beliefs about the conflict and the adversary and charting them across time, the analyst can note if changes are occurring—for example, one leader's beliefs are becoming more entrenched and less cognizant of opportunities to bargain. An analysis of this sort on the public speeches of Prime Minister Yitzhak Rabin of Israel and Pres. Anwar Sadat of Egypt following the 1973 Yom Kippur War suggests why peace was possible even though officially Egypt "lost" the war.[13]

An influential set of studies by Daniel Kahneman and Amos Tversky on the impact of policy makers' definitions of the situation on foreign-policy decision making became the basis for prospect theory.[14] In essence, their findings indicated that how individuals frame a situation shapes the nature of the decision they are likely to make. If policy makers perceive themselves in a domain of gains (things are going well), they are likely to be risk averse. But if their frame puts them in the domain of losses (things are going poorly), they are likely to be more risk prone or risk seeking. Decisions depend on how the policy maker frames the problem. Critical for determining whether a decision maker finds himself or herself in the domain of losses or gains is the individual's reference point or definition of the status quo. Problems arise when decision makers face situations where they perceive a discrepancy between what is happening and their reference point. The direction of the discrepancy indicates whether the decision maker interprets the situation as involving gains or losses, though decision makers appear to be more sensitive to discrepancies that are closer to their reference point than to those further away and, perhaps more importantly, to believe that "losing hurts more than a comparable gain pleases."[15] Interestingly, decision makers who are relatively insensitive to threats and constraints appear to have a penchant for the domain of losses and are risk seeking regardless of the situation, and similarly, decision makers who are generally anxious and responsive to what others think seek out the domain of gains and risk-averse behavior.[16] As Kahneman and Tversky observe, these individuals represent those who are willing to gamble no matter the odds or take out insurance no matter the cost.[17]

Experience also appears to count as an important influence on how policy makers interpret events.[18] With some expertise, they are more likely to rely on their knowledge and background and to engage situations on a case-by-case basis. Without expertise, decision makers are more affected by their personal predispositions, be they beliefs, motivations, or leadership style, or are led to depend on those whom they trust that have such experience. Moreover, policy makers feel more comfortable and confident dealing with domains in which they have some expertise and often drift toward those arenas.[19] Consider, for example, the effects of Dick Cheney on the foreign policy of the two Presidents Bush. Pres. George H. W. Bush had extensive experience in the foreign-policy-making process and could differentiate between relevant and irrelevant information as well as recognize inconsistencies in the information provided to him and exceptions to the rules—based on his own knowledge and expertise, he could say no to Cheney. His son, Pres. George W. Bush, came to office with little foreign policy experience or even international travel. By necessity he viewed Cheney as an expert and relied on Cheney's advice as well as his own beliefs regarding the importance of the United States and democracy in the world in making policy.

Often leaders and their perceptions are dismissed because they are viewed as constrained by the roles and institutions in which they find themselves. Prime ministers and presidents, for instance, are restricted by constitutions and norms to certain kinds of behavior. Even a head of government like Kim Jong Il of North Korea is constrained by his military as it is responsible for keeping him in power. But, interestingly, there is growing evidence that there are differences among leaders in whether they respect (work within) or challenge (go around) the constraints in their environments.[20] The two most recent presidents in Iran are a good example of this difference—compare Khatami, president until 2005, who, though charismatic, was concerned with working for change within the constraints of the political system in which he found himself, to Ahmadinejad, the current president, who pushes at the constraints in which he finds himself, willing to challenge the outside world and his own people.

Data suggest that leaders who are more pragmatic and opportunistic accept the constraints of their positions whereas the more ideological and strategic leaders tend to ignore the constraints unless they work to the leaders' advantage. Those who challenge constraints appear to engage in more confrontational behavior, commit the resources of their governments more readily, initiate activities, and take actions that involve or threaten to involve more than diplomacy. Leaders who challenge constraints often come to their positions with an agenda and seek "true believers" as advisors who will help them implement that agenda. They view themselves to be at the top of the decision-making ladder and are interested in controlling information flows; issues and events are not perceived as important or relevant unless they pertain to or affect the implementation of the agenda. On the other hand, leaders who respect constraints often seek out others' perspectives, are interested in diverse opinions, work well in a team, and focus on building consensus or working on a compromise. The problems that arise are those important to the constituents whose support is needed to stay in office. In effect, similar to the observations from research on experience, leaders who respect constraints are sensitive to the context and define as well as respond to problems on a case-by-case basis, while those who challenge constraints do so based on what they want or need—their personal predispositions.

Personal Predispositions Can Be Measured

A major reason researchers have offered for sticking with defining the problems facing policy makers from an outside observer's perspective revolves around the difficulties we have in assessing the subjective views of those involved in the process. Thus, the research to date that focuses on policy makers' own interpretations has tended to be case studies of individual incidents where it is feasible to interview some participants and secure access to archival materials. But within the last two decades, there has been a movement to engage in assessment at a distance—to use the recorded words and deeds of policy makers to infer their perceptions and interpretations. Only movie stars, hit rock groups, and athletes leave more traces of their behavior in the public arena than politicians. Indeed, few of a US president's or a British prime minister's movements or statements, for example, escape the media's and archivists' notice. With 24/7 coverage and the Internet, what leaders from around the world discuss is often beamed into our televisions and put onto the web. Such materials help us learn about how such figures are viewing and interpreting what is happening to them in more than a cursory fashion. Through content analysis, we can begin to develop images about these people and their ways of considering events even when they are essentially unavailable for the more usual assessment techniques. Content analysis does not require their cooperation, and researchers have begun to use this type of methodology to study what public figures are like and their interpretations of the problems they face.[21] In fact, there are now a number of software programs that assist in such analysis—for example, Profiler Plus.[22]

These content analysis techniques take advantage of the fact that communication is an important part of what political leaders do. Indeed, the web is full of the speeches, press conferences, and writings of political leaders. As governments seek to record what they are doing, the media captures their interactions among political leaders online, and political leaders themselves preserve their legacies. By using such materials, these assessment-at-a-distance techniques become unobtrusive ways of measuring how leaders view what is happening. Even though leaders may be shaping a communication for a specific audience or setting, we are able to take such intentions into account and learn about their effects by varying the kinds of material we study.

In effect, not only do the techniques—particularly those translated into software—make it feasible to construct a general profile of a particular leader or set of leaders, but also they make possible placing such profiles into perspective by examining contextual factors that indicate how stable the characteristics are with certain kinds of changes in the situation. We can ascertain what leaders are like in general, what kinds of information they are likely to respond to in the political environment, how they are likely to change their views with experience, and which situations are likely to be considered as involving gains versus losses, for example. With knowledge about both the general and contextualized profiles, the researcher and analyst gain a more complete portrait of a leader and the nature of how that person's rationality is likely to be bounded overall and in particular types of situations. It becomes possible to consider how policy makers are likely to interpret what is happening and the problems to be addressed instead of being limited only to how observers say that policy makers are likely to define situations.

Summary

Learning how policy makers view what is happening to them is critical to understanding how governments are likely to act. This statement is particularly true for those policy makers who have the authority to commit the resources of the government. Such policy makers' interpretations appear to be influenced by their beliefs, their experience, their thoughts about the chances of losing or winning in a situation, and their view of the constraints under which they operate. In effect, their rationality is bounded or conditioned by such characteristics. Depending on the nature of these characteristics, policy makers are likely to deal with situations on a case-by-case basis or to be directly and immediately influenced by their own personal predispositions. Researchers are developing techniques to assess these characteristics by taking advantage of the growing number of web-based speeches, interviews, and writings available on leaders from around the world. Some of these techniques have been turned into software programs that enable the analyst to examine all available materials and contextualize the resulting profiles.

Notes

(All notes appear in shortened form. For full details, see the appropriate entry in the bibliography.)

1. Snyder, Bruck, and Sapin, *Decision Making as an Approach*. See also Snyder et al., *Foreign Policy Decision-Making Revisited*.
2. Sylvan and Voss, *Problem Representation in Foreign Policy Decision Making*.
3. Beasley et al., "People and Processes in Foreign Policymaking."

4. Hermann et al., *Crisis-Prone Governments.*
5. Boin et al., *Politics of Crisis Management.*
6. Hermann and Kegley, "Rethinking Democracy and International Peace."
7. Hermann and Dayton, "Transboundary Crises through the Eyes of Policymakers."
8. Simon, "Human Nature in Politics"; and Simon, *Model of Bounded Rationality.*
9. Chollet and Goldgeier, "Scholarship of Decision Making," 157.
10. Simon, "Human Nature in Politics," 302.
11. Walker, "Role Identities and the Operational Codes of Political Leaders."
12. Schafer and Walker, *Beliefs and Leadership in World Politics.*
13. Walker, "Role Identities and the Operational Codes of Political Leaders."
14. Kahneman and Tversky, "Choices, Values, and Frames"; Kahneman and Tversky, "Prospect Theory"; and Tversky and Kahneman, "Advances in Prospect Theory." See also Farnham, *Avoiding Losses/Taking Risks*; and McDermott, *Risk-Taking in International Politics.*
15. McDermott, *Risk-Taking in International Politics*, 29.
16. Kowert and Hermann, "Who Takes Risks?"
17. Kahneman and Tversky, "Choices, Values, and Frames."
18. For example, see Preston, *President and His Inner Circle*; Beer, Healy, and Bourne, "Dynamic Decisions."
19. Stewart and Stasser, "Expert Role Assignment and Information Sampling"; Wittenbaum, Vaughn, and Stasser, "Coordination in Task-Performing Groups."
20. For example, see Keller, "Leadership Style"; Hermann, "Assessing Leadership Style"; and Hermann and Gerard 2009, "Contributions of Leadership."
21. For overviews of these techniques, see Young and Schafer, "Is There Method in Our Madness?"; Post, *Psychological Assessment of Political Leaders*; Schafer and Walker, *Beliefs and Leadership in World Politics*; and Hermann, "Using Content Analysis to Study Public Figures."
22. Social Science Automation Web site, http://www.socialscience.net/tech/overview.aspx.

Bibliography

Beasley, Ryan, Juliet Kaarbo, Charles F. Hermann, and Margaret G. Hermann. "People and Processes in Foreign Policymaking." *International Studies Review* 3, no. 2 (2001): 217–50.

Beer, Francis A., Alice F. Healy, and Lyle E. Bourne, Jr. "Dynamic Decisions: Experimental Reactions to War, Peace, and Terrorism." In *Advances in Political Psychology*, edited by Margaret G. Hermann, 139–68. London: Elsevier, 2004.

Boin, Arjen, Paul Hart, Eric Stern, and Bengt Sundelius. *The Politics of Crisis Management: Public Leadership under Pressure.* Cambridge: Cambridge University Press, 2005.

Chollet, Derek H., and James M. Goldgeier. "The Scholarship of Decision Making: Do We Know How We Decide?" In *Foreign Policy Decision-Making Revisited*, edited by Richard C. Snyder, H. W. Bruck, Burton Sapin, Valerie M. Hudson, Derek H. Chollet, and James M. Goldgeier, 153–80. New York: Palgrave Macmillan, 2002.

Farnham, Barbara, ed. *Avoiding Losses/Taking Risks: Prospect Theory in International Politics.* Ann Arbor, MI: University of Michigan Press, 1994.

George, Alexander L. "The 'Operational Code': A Neglected Approach to the Study of Political Leaders and Decision-Making." *International Studies Quarterly* 13, no. 2 (June 1969): 190–222.

Hermann, Margaret G. "Assessing Leadership Style: A Trait Analysis." In *The Psychological Assessment of Political Leaders*, edited by Jerrold Post, 178–212. Ann Arbor, MI: University of Michigan Press, 2005.

———. "Using Content Analysis to Study Public Figures." In *Qualitative Analysis in International Relations*, edited by Audie Klotz and Deepa Prakash. New York: Palgrave, 2008.

Hermann, Margaret G., and Bruce W. Dayton. "Transboundary Crises through the Eyes of Policymakers: Sense Making and Crisis Management." *Journal of Contingencies and Crisis Management* 17, no. 4 (December 2009): 233–41.

Hermann, Margaret G., Bruce W. Dayton, Matthew Smith, Havva Karakas-Keles, Azamat Sakiev, and Hanneke Dierksen. *Crisis-Prone Governments: A Study of 29 Pacific Rim Countries.* Syracuse, NY: Moynihan Institute of Global Affairs, 2009.

Hermann, Margaret G., and Catherine Gerard. "The Contributions of Leadership to the Movement from Violence to Incorporation." In *Conflict Transformation and Peacebuilding*, edited by Bruce W. Dayton and Louis Kriesberg, 30–44. New York: Routledge, 2009.

Hermann, Margaret G., and Charles W. Kegley. "Rethinking Democracy and International Peace: Perspectives from Political Psychology." *International Studies Quarterly* 39, no. 4 (December 1995): 511–33.

Kahneman, Daniel, and Amos Tversky. "Choices, Values, and Frames." *American Psychologist* 39, no. 4 (April 1984): 341–50.

———. "Prospect Theory: An Analysis of Decision under Risk." *Econometrica* 47, no. 2 (March 1979): 263–91.

Keller, Jonathan. "Leadership Style, Regime Type, and Foreign Policy Crisis Behavior: A Contingent Monadic Peace." *International Studies Quarterly* 49 (June 2005): 205–31.

Kowert, Paul A., and Margaret G. Hermann. "Who Takes Risks? Daring and Caution in Foreign Policy Making." *Journal of Conflict Resolution* 41, no. 5 (1997): 611–37.

McDermott, Rose. *Risk-Taking in International Politics: Prospect Theory in American Foreign Policy*. Ann Arbor, MI: University of Michigan Press, 2001.

Post, Jerrold, ed. *The Psychological Assessment of Political Leaders*. Ann Arbor, MI: University of Michigan Press, 2005.

Preston, Thomas. *The President and His Inner Circle: Leadership Style and the Advisory Process in Foreign Affairs*. New York: Columbia University Press, 2001.

Schafer, Mark, and Stephen G. Walker. *Beliefs and Leadership in World Politics: Methods and Applications of Operational Code Analysis*. New York: Palgrave Macmillan, 2006.

Simon, Herbert A. "Human Nature in Politics: The Dialogue of Psychology with Political Science." *American Political Science Review* 79, no. 2 (1985): 293–304.

———. *Model of Bounded Rationality*. Cambridge, MA: MIT Press, 1982.

Snyder, Richard C., H. W. Bruck, and Burton Sapin. *Decision Making as an Approach to the Study of International Politics*. Foreign Policy Analysis Project Series no. 3. Princeton, NJ: Princeton University Press, 1954.

Snyder, Richard C., H. W. Bruck, Burton Sapin, Valerie M. Hudson, Derek H. Chollet, and James M. Goldgeier. *Foreign Policy Decision-Making Revisited*. New York: Palgrave Macmillan, 2002.

Stewart, Dennis D., and Garold Stasser. "Expert Role Assignment and Information Sampling during Collective Recall and Decision Making." *Journal of Personality and Social Psychology* 69, no. 4 (October 1995): 619–28.

Sylvan, Donald A., and James F. Voss, eds. *Problem Representation in Foreign Policy Decision Making.* Cambridge, UK: Cambridge University Press, 1998.

Tversky, Amos, and Daniel Kahneman. "Advances in Prospect Theory: Cumulative Representation of Uncertainty." *Journal of Risk and Uncertainty* 5 (1992): 297–323.

Walker, Stephen G. "Role Identities and the Operational Codes of Political Leaders." In *Advances in Political Psychology*, edited by Margaret G. Hermann, 71–106. London: Elsevier, 2004.

Wittenbaum, Gwen M., Shannon I. Vaughn, and Garold Stasser. "Coordination in Task-Performing Groups." In *Social Psychological Applications to Social Issues.* Vol. 4, *Theory and Research on Small Groups*, edited by R. Scott Tindale, Linda Heath, John Edwards, Emil J. Posavic, Fred B. Bryant, Yolanda Suarez-Balcazar, Eaaron Henderson-King, and Judith Myers, 177–204. New York: Plenum Press, 1998.

Young, Michael D., and Mark Schafer. "Is There Method in Our Madness? Ways of Assessing Cognition in International Relations." *Mershon International Studies Review* 42 (1998): 62–96.

Chapter 7

Intent from an International Politics Perspective

Decision Makers, Intelligence Communities, and Assessment of the Adversary's Intentions

Keren Yarhi-Milo, PhD, Department of Politics and the Woodrow Wilson School of Public and International Affairs, Princeton University

Abstract: How do decision makers and intelligence communities infer the political and military intentions of the adversary? Using wide-ranging declassified primary documents from presidential archives, intelligence assessments, and interviews with US decision makers and intelligence analysts, I test three alternative indicators of intentions. These include the adversary's capabilities, strategic military doctrine, and behavior. The cases providing the evidence for these tests are the British assessments of Nazi Germany's intentions in the years leading up to the Second World War, US assessments of Soviet intentions under the administration of Pres. Jimmy Carter, and US assessments of Soviet intentions in the years leading to the end of the Cold War under the second administration of Pres. Ronald Reagan. I find that decision makers tend to view the world through the behavioral lens, placing importance on the actions of countries, including participating in binding international organizations, entering arms control agreements, and refraining from intervening in areas outside their legitimate sphere of influence. The US intelligence community, meanwhile, places greater emphasis on the capabilities of other nations, such as building up or scaling down their military forces, as indicators of intent. Intelligence products have limited success in influencing the decision calculus of decision makers, who rely on their own lens for understanding the adversary's intentions.

Introduction

One of the central tasks of statecraft is to predict the behavior of current or prospective adversaries. History teaches us, however, that this is perhaps one of the most challenging undertakings confronting decision makers (DM). Observing the dramatic changes that were taking place in Nazi Germany during the mid-1930s, members of the British cabinet argued fervently about the nature and scope of Hitler's ambitions. American DMs throughout the Cold War debated the objectives of the Soviet Union. Today, they debate the future intentions of a rising China and the political and military objectives of Iran.

International relations scholars have advanced propositions, implicit or explicit, about how states should gauge the intentions of other states. One school of thought in international relations argues that since the intentions of other states are difficult to discern with confidence, cautious DMs must assume the worst about their adversaries' intentions.[1] Other scholars argue that uncertainty about intentions should lead DMs to focus on the adversary's material power, especially its military and economic capabilities.[2] Yet others recognize the strong incentives for leaders to look beyond material indicators.[3] More recently, the literature has emphasized the importance of costly signals in revealing information about intentions.[4] The focus of this recent rationalist literature has been on the signaling of costly information about intentions; how perceivers extract and interpret such information has been almost entirely ignored.

In addition to theoretical approaches to discerning others' intentions, the evolution of adversarial relations throughout history also suggests important empirical questions: What prompts change in perceptions about intentions? What factors, for example, best explain the change in Pres. Ronald Reagan's beliefs about the intentions of the Soviet Union under Mikhail Gorbachev? Were these factors comparable to the ones that led members of Pres. Jimmy Carter's administration to reevaluate Soviet objectives in the late 1970s? On which type of evidence did British DMs focus when estimating the intentions of Nazi Germany in the 1930s? In all such cases, what role did intelligence estimates about the adversary's intentions play in shaping

DMs' inferences about the enemy's intentions? And were there systematic differences between the views of DMs and the views of the intelligence community providing advice to them?

Below I describe three competing theoretical perspectives on the links between intentions and changes in perceptions, which I term the capabilities thesis, the strategic military doctrine thesis, and the behavioral signals thesis. Then I summarize my research's methodology and main findings and briefly indicate the foreign policy significance of my findings.

Testing Alternative Explanations for Perceived Intentions

As a first step toward addressing these questions, I tested the causal relationship between three kinds of indicators and perceptions of an adversary's long-term foreign policy objectives and its intentions to use military force to achieve these objectives.[5] The first, called the capabilities thesis, posits that DMs infer the intentions of an adversary from different indexes of its military power. This thesis draws on several theories whose common denominator is the idea that, since governments are unlikely ever to be certain about others' intentions and because intentions are fluid and can easily change, prudent DMs should assume the worst about other states' intentions.[6] Thus, how aggressive an adversary could be is essentially a function of how powerful it is. Accordingly, this thesis predicts that DMs will rely heavily on assessments of an adversary's capabilities to infer its intentions. For example, a country is likely to be seen as signaling hostile intentions if it devotes more resources to building up its military capabilities and develops and deploys offensive capabilities.

The second thesis, which I term the strategic military doctrine thesis, posits that DMs infer intentions of the adversary from the adversary's military doctrine. Building on the logic of the offense-defense theory, scholars have pointed out that, by providing a set of ideas about how to employ the instruments of military power, a state's military doctrine is likely to be seen as a valuable indicator about the objectives of others.[7] Accordingly, adopting offensive conventional or nuclear doctrines is expected to increase the probability that others will perceive that country as having hostile intentions. Conversely, adopting deterrent or defensive doctrines, either conventional or nuclear, is expected to increase the probability that others will infer benign intention.

The third theoretical perspective, termed the behavioral signals thesis, posits that certain kinds of costly actions are particularly useful in revealing information about the objectives of the adversary. The thesis is rooted in various strands of work in international relations—the democratic peace theory, neoliberal institutionalism, social constructivism, and the rationalist literature on signaling. According to this thesis, actions such as joining binding international institutions, initiating domestic democratic reforms, embarking on military interventions, and signing arms control agreements can be perceived by others as providing costly, reassuring signals of benign intentions. Alternatively, withdrawing from such institutions and commitments is likely to lead others to infer malign intentions.

Methodology

To test the three theses about perceived intentions, I employed the historical case-study method. I examined the British assessments of Nazi Germany's intentions in the years leading up to the Second World War, US assessments of Soviet intentions under the Carter administration, and US assessments of Soviet intentions in the years leading to the end of the Cold War under the second Reagan administration. There are several ways we can examine how well each of the three theses explaining perceived intentions fares against the empirical record. To test each one in the case studies, I examine three aspects of decision making. First, relying on private statements by DMs and intelligence agencies, I look at covariation between perceptions of the adversary's capabilities, strategic military doctrine, or behavior and shifts in perceptions of its political and military intentions. I also examine whether the direction of change in perceived intentions accords with the predictions of each thesis. Second, I examine the reasoning invoked to support assessments of intentions. Here I look at the explanations provided in private statements and evaluate whether these explanations were informed by the factors emphasized in each of the three theses. For example, if the capabilities thesis is correct, we should expect DMs to link their assessments of the adversary's intentions to indicators associated with the adversary's military capabilities. Third, I look for covariation between the individual DM's statements about his perceptions of the adversary's intentions and the policies towards the adversary that he advocates.

I analyze perceptions about the adversary's intentions among two groups—the intelligence community and key DMs. My research is primarily based on over 20,000 primary documents that I collected from the Reagan and Carter presidential libraries, the National Security Archive, National Archives, the British National Archives, and the Israel Defense Forces Archives in Israel. Additionally, to date I have conducted over two dozen interviews with American and Israeli DMs and retired intelligence analysts. For assessments of the US intelligence community during the Cold War, I use the declassified national intelligence estimates (NIE) on the Soviet Union as well as research papers, memoranda of conversations, and documents produced by the Office of National Estimates in the CIA. The community regularly assessed Soviet intentions in the 11-4 and to a lesser extent the 11-8 series of NIEs, and these were supplemented by occasional special NIEs (SNIE). For key US DMs' views, I make extensive use of declassified archival documents such as protocols of National Security Council meetings, national security decision directives, presidential directives, memoranda of conversations between the president and other important DMs in his administration, and additional documents, telegrams, and protocols of meetings where DMs discussed the threat posed by the adversary. For assessments of the British intelligence community prior to World War II, I review reports produced by the intelligence staffs of the War Office, the Air Ministry, the Admiralty, and the Foreign Office, as well as reports and protocols of meetings of the Defence Requirements Sub-Committee and the Chief of Staff Sub-Committee. To analyze the views of British DMs, I look at the protocols of Cabinet meetings, Cabinet memoranda, Foreign Office documents, and protocols and memoranda of various Cabinet committees, such as the Committee for Foreign Policy, the Defence Requirement Committee, and various ad hoc committees.

Findings

How did policy makers—specifically, civilian DMs and intelligence communities—infer the political and military intentions of their adversary during these historical episodes? When and why do these beliefs about intentions change? My research has yielded some surprising and novel findings. The evidence I examined reveals that intelligence communities and civilian DMs use quite different analytic lenses to assess intentions.

Specifically, in all three cases, the evidence from civilian DMs' assessments of the adversary's political intentions lent the strongest support to the behavioral signals thesis and rather weak support to the capabilities and strategic military doctrine theses. The timing and direction of changes in perceptions of intentions and the reasoning invoked by DMs to support their assessments, as well as the policies they advocated, were congruent with the hypotheses inferred from the behavioral signals thesis. American DMs in two very different administrations, as well as British DMs in an entirely different historical era, all privileged indicators associated with the recent behavior of their adversaries, rather than the adversaries' capabilities, when assessing the nature and scope of their foreign policy goals.

What types of behavioral signals seem to have mattered the most? The historical evidence does not indicate one particular behavioral signal that was most important. In the late 1970s, Soviet interventions in the Third World exerted the most significant effects upon the perceptions of intentions among members of the Carter administration. However, in the second term of the Reagan administration, Gorbachev's domestic behavior toward dissidents, his 1987 signing of the Intermediate-Range Nuclear Forces (INF) Treaty, and his withdrawal of Soviet forces from Afghanistan combined to change US perceptions of Soviet intentions. For British DMs in the 1930s, Hitler's militarization of German society, the annexation of Austria, and, most importantly, the invasion of Czechoslovakia ultimately led Chamberlain and members of his cabinet to change their view of the scope of Hitler's revisionist intentions.

In each of the three cases, DMs used a similar process of interpreting and reacting to their adversary's actions. Prima facie, it looks as if one particular action was responsible for a sudden shift in perceptions of the adversary. The Soviet invasion of Afghanistan, for example, has often been cited as the single most important action to cause a shift in President Carter's perceptions of the Soviet Union. But the evidence indicates that Carter's perceptions had begun to change as early as 1978, when the Soviets intervened in the Horn of Africa. Afghanistan was the tipping point. Likewise, in the British case, the evidence indicates that British perceptions of Germany's intentions had begun to shift in mid-December 1938, following intelligence reports of a possible German attack in Western Europe. The invasion of Czechoslovakia in March 1939 merely reinforced British DMs' confidence that Germany's intentions were more grandiose than they had thought.

Perceptions of intentions did not change to the same degree among all DMs. Thus, Secretary of State Cyrus Vance, Secretary of Defense Caspar Weinberger, and the British ambassador in Berlin, Sir Nevile Henderson, did not significantly change their perceptions of the intentions of the key adversaries their countries faced. Their perceptions certainly did not change to the same extent as those of their leaders—Carter, Reagan, and Chamberlain. Preexisting beliefs, and the degree to which they were entrenched, probably played a role in preventing more radical transformations of perceived intentions. Yet even those DMs who never "updated" their beliefs in light of new evidence consistently used the adversary's recent behavior to support assessments about their intentions. To Vance, the invasion of Afghanistan indicated that the Soviet Union was still an opportunistic power, but not an expansionist one. To Weinberger, the withdrawal from Afghanistan came only because the Soviets had recognized they could not win that war. To Henderson, the German invasion of Prague was still compatible with the long-held conviction that Germany's intentions were limited.

American and British DMs were also similar insofar as they did not use certain indicators to infer intentions. During the mid-to-late 1970s, the Soviets built up their military capabilities, especially their nuclear forces, to such an extent that some described it as a push to gain superiority in the arms race. During the 1980s, Soviet capabilities declined. British DMs in the 1930s were aware of the impressive rate, scale, and magnitude of Germany's rearmament efforts. Yet the historical documents do not suggest that these changing capabilities fundamentally altered American or British perceptions of their adversaries' intentions. Empirical support for the capabilities thesis is weak: the turning points in DMs' perceptions of intentions were at best only partly consistent with changes in the adversary's capabilities. In private discussions and writings about the adversary, DMs only rarely linked their assessments of the adversary's capabilities to their interpretations of its foreign policy goals. It is true that US-Soviet détente collapsed at about the same time the Soviets were building up their nuclear forces, the Cold War ended at about the same time the Soviets scaled down their conventional forces, and World War II erupted following a massive German military expansion. But the evidence I have presented in the empirical cases suggests that the correlation should not be mistaken for a causal relationship. During each of these historical periods, observations of the adversary's behavior, rather than observations about its changing military capabilities, produced a change in perceptions of its intentions and resulted in changed foreign and defense policies.

Nevertheless, changes in the adversary's capabilities were significant in three ways. First, these changes often led DMs to raise questions about the adversary's political objectives and military intentions. Second, concern about the extent of the changes, as well as the types of weapon systems being developed and deployed, was critical in shaping the ways countries responded in their foreign policy and defense planning. Third, capabilities were the most significant indicators upon which the United States and British intelligence communities based their perceptions of intentions.

Unlike DMs, when gauging the adversary's intentions, intelligence communities gave greater weight to capabilities over behavior and doctrine. The records of the available intelligence estimates about the Soviet threat during the Cold War are riddled with assumptions about the causal relations between Soviet capabilities and their likely foreign policy behavior in the future. The NIEs of the 1970s and 1980s repeated the claim that Soviet decisions about how, where, and when to expand would be driven by Soviet assessments of the correlation of forces between the two superpowers. Similarly, British intelligence reports of Germany's foreign policy suggested an intimate link between capabilities and intentions, claiming that Germany's foreign policy would be dictated by calculations about when the German army would reach its peak efficiency and numerical superiority. Additionally, both the British prewar and the American Cold War intelligence communities appear to have dedicated only a small portion of their efforts to analyzing the political intentions of the adversary. In the 1970s, the vast majority of NIEs analyzed many different aspects of the Soviet strategic forces, but the sections that analyzed Soviet intentions and strategic objectives were remarkably underdeveloped. The intelligence community's discussion of Soviet intentions improved somewhat during the mid-1980s, when it struggled to understand Gorbachev's motivations and ultimate objectives. The low priority of political intentions was also evident in British intelligence assessments of the 1930s, which contained very little analysis about the scope and nature of Germany's long-term foreign policy goals under Hitler.

In all three of the cases discussed here, the analysis of intentions by intelligence agencies suffered from similar biases and fallacies, which were later recognized by scholars. For instance, as shown in the case studies, mirror imaging—attributing to the subject one's own values, perceptions, and behavior—led to inaccurate estimates about the adversary's military capabilities, which consequently affected assessments about intentions. Also, in all three cases, assessments fluctuated between over-estimating and underestimating the adversary's capabilities. British intelligence moved from underestimating the German military in the early to mid-1930s to exaggerating its abilities between mid-1936 and early 1939. Similarly, the US intelligence community shifted from an underestimation of Soviet strategic forces to an overestimation during the 1970s and 1980s.

How much influence did intelligence assessments of the adversary's political intentions have on the perceptions of civilian DMs? The short answer is that their effect was marginal, at best. National DMs appear to have reached conclusions about the adversary's foreign policy objectives independent of intelligence estimates. To a large extent, they preferred to make these general political judgments for themselves. Indeed, Robert Vansittart, Zbigniew Brzezinski, and George Shultz were all critical of intelligence estimates, believing that they had better knowledge of the adversary and could therefore offer better insights about the adversary's foreign policy objectives. This notion is nicely captured by Vansittart's criticism of British intelligence reports on Germany's military power: "Prophecy is largely a matter of insight. I do not think the Service Departments have enough. On the other hand they might say that I have too much. The answer is that I know the Germans better."[8] The evidence presented in research shows that intelligence assessments were not congruent with the opinions and assessments of key civilian DMs at the time. But perhaps most important, there is no evidence to suggest that intelligence reports—even when DMs read them—led them to reconsider their own assessments.

These findings echo the conclusions of scholars who have examined the correlation between intelligence and foreign policy in other cases. For example, after analyzing US intelligence views of the threats from Japan and Germany, David Kahn concludes, "Intelligence had little to do with American assessments of Germany and Japan before December 1941."[9] Similarly, interviews I conducted with Maj Gen Shlomo Gazit, retired, head of the Israel Defense Forces intelligence branch in the mid-1970s, suggest that Israeli DMs paid little attention to the Israeli intelligence about Anwar Sadat's political intentions in the period following the 1973 Arab-Israeli War. Moreover, the analysis here has shown that intelligence reports relied on different indicators than those used by DMs and reached different conclusions about the adversary's foreign policy objectives. Both the timing and the reasoning provided by the intelligence community for its reassessments of the adversary's intentions were markedly different from those of the key civilian DMs. As a result, key civilian DMs viewed intelligence assessments on issues pertaining to the adversary's political intentions as largely irrelevant.

So far I have discussed the relevance of the adversary's behavior and capabilities for assessments of intentions. The third category of signals examined in this study is the adversary's strategic military doctrine. DMs rarely invoked the adversary's doctrine in their deliberations about its political intentions, while intelligence communities had trouble clearly defining the adversary's military doctrine. During the 1970s, the US intelligence community was split in its assessments of Soviet nuclear doctrine and the intentions to be inferred from it. During the 1980s, the US intelligence community was slow to recognize changes in Soviet military doctrine and was split in its evaluation of what these changes signified about Gorbachev's intentions. During the 1930s, British intelligence recognized the German strategic military doctrine as offensive, but it failed to identify the particular characteristics of the blitzkrieg and the role of the Luftwaffe in planned military operations. In all three cases, then, the adversary's strategic military doctrine did not serve as a straightforward indicator of intentions, and its role in the assessment of political intentions was marginal. The adversary's strategic military doctrine did play a role in the American assessments of Soviet military intentions for both Cold War cases, but even then, it was secondary to assessments of Soviet capabilities.

Implications

This study has several practical policy implications. First, and perhaps most obviously, a state's (non–capability based) behavior is likely to have the greatest effect on how others will judge its intentions. This means that states can most effectively signal benign intentions by joining binding international organizations, entering arms control agreements, and refraining from intervening in areas outside their legitimate sphere of influence. However, even through such behavior, reassurance is difficult, and the historical record suggests uncertainty about the extent to which these actions will have the desired effect. Cognitive biases and preexisting beliefs are likely to taint others' interpretation of these actions and affect the resulting perceptions of intentions. Further, the cases I have examined suggest that the pattern of change in perceived intentions appears to mirror one of punctuated equilibrium. Yet a close look at the evolution of perceived intentions indicates that the accumulation of several actions, rather than one particular costly action, produces fundamental change in perceptions. This means that policy makers should be aware that a single reassuring act is unlikely to decisively change others' assessments. Similarly, behavior that might give others reason to be concerned, if followed by a reassuring action, is unlikely to lead them to reach firm and immutable conclusions about one's intentions.

Second, while behavior appears to matter most, building up or scaling down one's military forces is unlikely to change how other DMs perceive one's political intentions. Prudent DMs should not, however, conclude that this means they need not worry about the consequences of a large-scale military buildup. The historical evidence indicates that armament efforts by the adversary did lead observers to raise, though not answer, questions about the adversary's objectives.

A third set of policy implications pertains to the quality of analysis generated by the intelligence community about adversaries' political and military intentions. This study was not designed to analyze intelligence failures, and the question of whether the intelligence community got it right or not is purposefully left unanswered. As a result, I am in no position to offer concrete recommendations as to whether and how the intelligence community in the United States or elsewhere needs to be transformed. This study, however, does point to practices in imputing intentions that may have prevented the intelligence community from reaching accurate conclusions about the adversary's foreign policy goals. Most relevant in this context are the following observations.

First, the NIEs did not state the assumptions underlying the intelligence community's estimates of the adversary's political intentions. Thus, it is unclear whether the intelligence community was even aware of its (repeated) practice of discerning intentions from capabilities and whether it recognized at all that this premise was decisive in driving estimates about Soviet foreign policy intentions. Second, the premise that the perceived correlation of forces would determine Soviet foreign policy should have been examined as to whether it relied on direct evidence or on assumptions. The intelligence community should have asked what an alternative picture of Soviet intentions would look like if that premise were removed from the equation. The NIEs, however, contained no alternative explanations for the adversary's behavior and consequently did not even entertain the possibility that the intentions attributed to the adversary based on capabilities might be imprecise, if not altogether wrong. Questioning this premise might have allowed the US intelligence community to consider the possibility that Gorbachev's intentions were different from those of earlier Soviet leaders. Yet, to be fair, even if the practice of inferring intentions had been conducted with more rigor and imagination, there is no guarantee that the intelligence community would have produced more accurate assessments of Soviet intentions. Nor is there reason to believe that the sorts of improvements I'm suggesting in the assessments of adversary intentions would have led civilian DMs to pay more attention to the views of the intelligence community. Nonetheless, if no effort is made to institutionalize practices designed to counteract the bias this study has identified, which privileges capabilities over behavior, then one should expect it will continue to affect the intelligence community's assessments of adversaries' intentions.

Notes

(All notes appear in shortened form. For full details, see the appropriate entry in the bibliography.)

1. Mearsheimer, *Tragedy of Great Power Politics.*
2. Waltz, *Theory of International Politics.*
3. Walt, *Origins of Alliances.*
4. Fearon, "Signaling Foreign Policy Interests"; and Kydd, *Trust and Mistrust in International Relations.*
5. Yarhi-Milo, "Knowing Thy Adversary."
6. Mearsheimer, *Tragedy of Great Power Politics.*
7. Posen, *Sources of Military Doctrine*; and Glaser, "Political Consequences of Military Strategy."
8. Watt, "British Intelligence."
9. Kahn, "United States Views of Germany and Japan," 476.

Bibliography

Betts, Richard. *Enemies of Intelligence: Knowledge and Power in American National Security.* New York: Columbia University Press, 2007.

Edelstein, David. "Managing Uncertainty: Beliefs about Intentions and the Rise of Great Powers." *Security Studies* 12, no. 1 (Autumn 2002): 1–40.

Fearon, James. "Signaling Foreign Policy Interests: Tying Hands versus Sinking Costs." *Journal of Conflict Resolution* 41, no. 1 (1997): 68–90.

Glaser, Charles L. "Political Consequences of Military Strategy: Expanding and Refining the Spiral and Deterrence Models." *World Politics* 44, no. 4 (July 1992): 497–538.

Jervis, Robert. "Cooperation under the Security Dilemma." *World Politics* 30, no. 2 (January 1978): 167–214.

———. *The Logic of Images in International Relations*. Princeton, NJ: Princeton University Press, 1970.

Kahn, David. "United States Views of Germany and Japan in 1941." In *Knowing One's Enemies: Intelligence Assessment before the Two World Wars*, edited by Ernest R. May, 476–501. Princeton, NJ: Princeton University Press, 1984.

Kydd, Andrew. *Trust and Mistrust in International Relations*. Princeton, NJ: Princeton University Press, 2005.

May, Ernest R., ed. *Knowing One's Enemies: Intelligence Assessment before the Two World Wars*. Princeton, NJ: Princeton University Press, 1984.

Mearsheimer, John J. *The Tragedy of Great Power Politics*. New York: Norton, 2001.

Posen, Barry. *The Sources of Military Doctrine: France, Britain, and Germany between the World Wars*. Ithaca, NY: Cornell University Press, 1984.

Walt, Stephen. *The Origins of Alliances*. Ithaca, NY: Cornell University Press, 1987.

Waltz, Kenneth. *Theory of International Politics*. New York: Random House, 1979.

Watt, Donald. "British Intelligence and the Coming of the Second World War in Europe." In *Knowing One's Enemies: Intelligence Assessment before the Two World Wars*, edited by Ernest R. May, 237–70. Princeton, NJ: Princeton University Press, 1984.

Yarhi-Milo, Keren. "Knowing Thy Adversary: Assessments of Intentions in International Politics." PhD diss., University of Pennsylvania, 2009.

Chapter 8

Social Cognitive Neuroscience

The Neuroscience of Intent

Sabrina J. Pagano, PhD, NSI, Inc.

Abstract: While a variety of scholars and practitioners across several disciplines have addressed the intent question, the present chapter will consider the ways in which the relatively new field of social cognitive neuroscience (SCN) can illuminate our understanding of others' intent. The field of SCN may yield greater insight into intent by allowing better understanding of the processes being engaged in response to psychological events or interaction. We thus may best engage with, change, or counteract those processes. However, it is likely that the best understanding of how to decipher intent, as in the study of other complex problems, will be a multimethod approach rather than one focusing solely on SCN. While no method should be viewed as a magic wand or panacea, SCN can contribute to our understanding of intent by increasing insight into the way we make attributions for others' intentions. SCN offers the promise of examining how this process might differ when we make more or less accurate assessments of intent and thus may help us understand how we might do so more effectively. With time and continued work in this area, this increased insight might extend into an improved understanding of how to measure and model others' intent and not solely our interpretation of it.

Introduction and Methodological Overview

In its simplest form, intent can be thought of as a determination to engage in a given action with the goal of achieving a specific outcome.[1] Deciphering the intentions of others, however, is far from a simple task. Complicating the process of deciphering intent is the potentially complex relationship among people's attitudes, intent to act, and their actual behavior.[2] Arguably, the first step in this process involves acknowledging that others may have mental states (i.e., intentions, beliefs, emotions, and desires) that differ from our own. Beyond merely understanding that others may think and feel differently than we do, a true understanding of others' intent also requires the ability to determine the nature of those feelings and thoughts, as well as their likely associated behaviors. This ability has been referred to as mentalizing or theory of mind.[3]

While a variety of scholars and practitioners across several disciplines have addressed the intent question, the present chapter will consider the ways in which the relatively new field of social cognitive neuroscience (hereafter SCN) can illuminate our understanding of others' intent. First, however, a brief description of one of the most promising brain measurement techniques—and its strengths and weaknesses—is warranted.

Measuring Brain Activity: Functional Magnetic Resonance Imaging

At the outset, it is critical to dispel the common misperception that there is a straightforward, one-to-one relationship between neural firing in a given region of the brain and particular tasks or cognitive processes. For example, it would be overly simplistic to refer to the "aggression spot" of the brain. The more complex the macro-level process, the more likely that a distributed network of brain elements is involved.[4] Moreover, it is a challenge to ascertain with any certainty whether a particular brain structure or network is involved solely with one process; indeed, available evidence appears to suggest otherwise.[5] That said, there are conditions under which we can reasonably expect to observe a similar pattern of activation when people perform a given task if that pattern has been observed previously. This occurs when a particular *set* of brain regions is consistently activated in response to a particular type of task over dozens of studies examining multiple individuals and that pattern of activation is *not* observed during other activities. Thus, as a shorthand, we may find it useful to speak of distributed networks being associated with specific cognitive processes. As Cacioppo and his colleagues caution, however, even seemingly clear patterns of activation do not eliminate the possibility that the networks being activated are "part of a sufficient but not a necessary neural mechanism for [a given] information-processing operation."[6]

SCN allows us to examine social phenomena and processes using the tools traditionally utilized by cognitive neuroscience, such as neuroimaging.[7] Since another chapter in this volume is dedicated to neuroimaging methods, their treatment here will be brief. The most popular of the neuroimaging techniques, and the principal approach reported in this chapter, is functional magnetic resonance imaging (fMRI). A majority of fMRI studies use what is known as blood-oxygen-level dependent (BOLD) contrast to determine which regions of the brain are more or less active during a psychological task. BOLD is based on two presuppositions: (1) blood flowing to more active regions of the brain is higher in oxygen content than blood in less active brain regions; and (2) oxygenated blood has different magnetic properties than deoxygenated blood. An MRI works by mapping the varying magnetic signals to determine the pattern of blood flow in the brain.

Like any other methodology, fMRI (and neuroimaging in general) has strengths and weaknesses that must be considered when deciphering the degree to which it can facilitate our understanding of intent. One major strength of fMRI is that the precision with which it can measure the brain's activity (i.e., its temporal and spatial resolution) is greater than that of other brain imaging techniques like positron emission tomography (PET). This precision allows researchers to examine a greater number and type of questions. Another strength of fMRI is the potential to discriminate between the cognitive processes that individuals appear to use when engaging in differing tasks, such as deciding to pursue one course of action versus another.[8]

This information goes beyond what we are able to decipher, for example, by simply observing behavioral outcomes (e.g., that someone pursues course of action A versus course of action B).

A major limitation of fMRI is that to obtain measurements of which areas of the brain are engaged during certain tasks, the individual under study must be exposed to the task or stimulus repeatedly, potentially leading to response contamination when individuals become accustomed to the task and no longer react in an unbiased way.[9] Second, because of the way in which fMRI data must be collected, comparisons between groups—so often of interest when examining social science questions—are atypical. One reason for this limitation is that the scanner environment itself is quite sterile. Those undergoing an fMRI must remain completely still inside a very noisy scanner; they cannot speak, and face-to-face interactions cannot occur for the duration of the brain imaging,[10] which typically lasts between 30 and 60 minutes. People being scanned wear video goggles that present stimuli to which they indicate responses on a button box typically composed of two to five buttons. This is hardly the rich social environment in which people interact daily. However, cutting-edge research techniques are beginning to mitigate some of these limitations. For example, protocols have been established that allow an examination of real social interactions (albeit still in the scanner context)[11] or of the neural activity of two people as a coupled system.[12]

Intent-Related Research in Social Cognitive Neuroscience

People frequently make inferences about others' qualities and mental states based on relatively impoverished inputs, a process known as thin slicing.[13] For example, in one study, students made judgments about a teacher's nonverbal behavior by viewing less than 30 seconds of a silent video of the instructor teaching.[14] These quick judgments were essentially the same as those made by students who had taken the instructor's class for a semester. This type of result has been replicated in a laboratory versus more natural settings and across modes of expression (e.g., tone of voice versus facial cues). Nonetheless, several variables can moderate (i.e., weaken or strengthen the relationship between the input and the output of) thin slicing. These include the type of target being judged (e.g., more extroverted and more expressive people are easier to judge), the person judging the behavior (e.g., less dogmatic and more cognitively complex people do better), the culture of the judges and targets (people are typically better at assessing targets from their own culture), and context (namely, judgments improve when the context is appropriate and diagnostic, such as observing and diagnosing someone as an extrovert in a large group versus an intimate social setting).[15] Thus, despite the amazing potential implicated by our ability to thin slice, we often appear to fall short of easily and reliably understanding others' intent. The very challenges implied by screening lines at airports, elaborate job application protocols, and the like reveal that we are still working on the intent puzzle. Fortunately, SCN can contribute some of the pieces.

Broadly speaking, much of the SCN literature relevant to the question of intent focuses on the ability of a social observer to determine the intentions of another individual. Less research appears focused on determining the intentions of a given actor and the cohesion—or lack thereof—among the actor's attitudes, intentions, and behaviors. The following is a brief overview of the most relevant findings in several categories of study related to the perception and understanding of others' intent.

Measuring Our Ability to Decipher Others' Intent

Theory of Mind. As discussed above, theory of mind (ToM) refers to the ability of people to understand that others have mental states that may differ from their own and to infer what those mental states might be.[16] Some researchers have gone so far as to say that ToM includes the ability to explain and predict the behavior of others of the same species.[17] In general, a successful analyst would likely perform well on ToM tasks. However, measuring ToM aptitude does not require SCN—it can be determined using simple paper-and-pencil tests. What cannot be determined easily from these written tests, however, are the subprocesses in which people engage when they are more or less successful at ToM. SCN in time can provide us with this kind of information by differentiating the cognitive processes (e.g., making judgments about familiar or unfamiliar others, inferring temporary versus stable states or dispositions) involved in the selective engagement of different brain networks.

For example, a general consensus in the neuroscience literature has identified the medial prefrontal cortex (mPFC) as a brain region that is crucial to social understanding, such as attempts to decipher the comparatively stable dispositions (e.g., personality traits such as extroversion or conscientiousness) of others.[18] Additional regions, such as the temporal parietal junction (TPJ), may be recruited to infer transient states such as the goals, intentions, and desires of other people.[19] Similarly, different regions of the brain are engaged when making judgments about familiar versus unfamiliar others or when specifically visual (as opposed to other media) representations of target behavior are presented to the perceiver.[20] More importantly, brain regions may be differentially engaged depending on the nature of the intention. For example, Ciaramidaro and colleagues propose and provide evidence for distinct brain activity based on whether an individual's goal or intention is social or private and whether an interaction is to occur in the present or the future.[21]

To the extent that SCN will be able to give us confident insights into the subprocesses involved in performing various tasks, we can train analysts to engage in these processes to achieve a better understanding of others. One method of training might be to use neurofeedback, which appears to operate based on operant conditioning (wherein positive reinforcement can be used to continually modify behavior so that it becomes successively closer to the desired outcome). Neurofeedback enables people to monitor in real time, and in some cases modify, biological states such as brain activity. This ability is useful to the extent that one is trying to develop or train a particular neural or other response. A full review of the neurofeedback literature is beyond the scope of the present paper; however, the success of this method across several domains (e.g., treating attention deficit disorder or migraines) suggests that it might be a worthwhile area for future investigation. Additionally, to the extent that people successfully engaging in various kinds of ToM tend to demonstrate a particular kind and degree of neural activation for each kind of task, we could potentially use fMRI measurement to select people who perform above a desired threshold. However, this suggestion comes with the caveat that this assessment—which necessarily has some measurement error—should be used as part of a battery-style screening tool and not as a stand-alone criterion for selection or dismissal.

Emotion Recognition. In making attributions about others' intentions, we often look at facial expressions and other nonverbal indicators for clues to their emotional states. Recent research suggests that relatively independent brain systems may process different kinds of basic emotions (i.e., biologically based emotions that are thought to be cross-culturally universal, such as anger or disgust).[22] At the same time, some structures or networks appear to be variously involved in the processing of any stimulus of emotional significance. Neuroscience researchers have focused mainly on the amygdala as a central brain structure of interest in exploring what happens as we engage in the emotion recognition process,[23] and there is evidence that the amygdala in fact allows us to understand and process the emotional significance of incoming stimuli in a relatively automatic way.[24] For example, selective damage to the amygdala results in deficits to an individual's ability to recognize fear in others.[25] However, the amygdala is not the only region implicated in the ability to detect emotion from facial cues. Other regions of the brain (e.g., the ventromedial prefrontal cortex, insula, and basal ganglia) appear to be implicated in people's ability to identify other facial expressions, such as disgust.[26] As the body of SCN work on perceived hostility and disgust grows, we may discover a promising source for identifying individuals who can assess potentially dangerous intentions. We could operationalize this knowledge in the future by training or selecting analysts in the same way as described in our earlier discussion of ToM. The growing SCN literatures on attributions of trustworthiness and on trust interactions may inform training and selection from a complementary perspective.

Action Observation. Another promising area for understanding intent is the study of action observation, which arises when people perceive biological motion (i.e., motion that is consistent with the biomechanics of living organisms) that suggests an intentional and goal-directed action is being enacted.

As discussed by Lieberman, the majority of action observation studies to date have focused on brain activity related to reaching to grasp or associated hand movements. Importantly, as Lieberman goes on to discuss, the regions of the brain that become active in response to action observation (compared with control stimuli) tend to become increasingly active when observing action that is specifically functional (e.g., grasping a coffee mug right side up versus upside down).[27] One potential benefit of SCN studies of action observation to our understanding of intent is as part of a selection or training tool aimed at optimally identifying behaviors associated with hostile intention (e.g., reaching for and grasping a gun). As action observation studies expand their repertoire of observed activities, they may yield greater rewards in understanding intent.

Measuring Others' Intent: Initial Research

Intent is an important variable to examine because it may be a precursor to actual behavior. Attempts to convert people's hostile intentions through persuasion are aimed ultimately at decreasing the corresponding hostile behaviors. To the extent that we can better understand how to influence people's intentions, we can thus address a core concern for analysts. In our attempts to convert people's hostile intentions through persuasion, it is important to understand what might predict both their intentions in response to a received persuasive message and their later behavior. The first step along this path comes from understanding what happens when people are successfully persuaded by a message to which they are exposed.

Falk and her colleagues undertook an investigation of the brain regions involved when people are successfully versus unsuccessfully persuaded.[28] During successful persuasion (and not during unsuccessful persuasion), active brain regions included those that together have been consistently shown to be involved in social cognition or mentalizing.[29] These include the dorsomedial prefrontal cortex (dmPFC), bilateral posterior superior temporal sulcus (pSTS), and the bilateral temporal poles. As in other kinds of research, the absence of other likely explanatory mechanisms (e.g., activation of the reasoning and working memory networks) in the Falk et al. study is noteworthy since it strengthens the narrative positing that specifically social cognitive and mentalizing processes are invoked during successful persuasion. More importantly, these findings were replicated across two distinct cultural groups and two forms of media, lending further confidence to these findings and potentially extending the degree to which they might be generalized.[30] Continued replication of this pattern of results would enable us to gain a better understanding of people's likely intentions and subsequent behaviors by examining their neural responses to a persuasive message, without having to directly ask them (though this capability is still a long-term goal).[31] Future SCN work should examine the precise relationship between successful persuasion and people's downstream intentions and/or behaviors; to date, there is no published body of work in this area. While there might not be a perfect concordance between people's intentions to act and their subsequent behaviors, it is safe to say that being in a specific intentional state increases the likelihood that someone will engage in intention-consistent behavior.[32]

The caveat remains that an examination of a single person's brain activity should not be taken as a direct indicator of subsequent behavior. In interpreting these findings, it also is important to note that this work used a topic for which people did not hold strong prior attitudes. As with other preliminary findings reported here, several replications and boundary tests for the limits of the initial effects or relationships are required before any conclusions might be drawn. Moreover, the act of measuring attitudes or intent can under some conditions alter later behavior, a process that in the marketing literature is known as the mere measurement effect. While the consideration of intent certainly is more complex than product selection, we might expect that some of the processes engendered by measurement may similarly alter the course of events to come. Despite these caveats, these initial findings extend our understanding of the processes invoked during successful persuasion, which may similarly predict intentions and downstream behaviors. Keeping abreast of related findings as they accumulate will allow us to determine the degree to which these preliminary findings can or cannot be generalized.

Case Study: How SCN Can Improve Understanding

To understand how SCN contributes to an improved understanding of various phenomena, let us examine a historical debate within social psychology on the motivational and behavioral implications of feeling different emotions in response to another's suffering. Several years ago, Daniel Batson, a prominent social psychologist, set about examining the relationship between empathy and altruistic behavior in the face of another's suffering.[33] His goal was to demonstrate that feeling empathy toward a suffering individual was distinguishable from feeling distress. Batson defined empathy as an emotional response congruent with the perceived welfare of another. When the other person is suffering, empathy may be composed of other-oriented feelings of tenderness, compassion, and similar affects. Personal distress, on the other hand, is a self-focused emotion characterized by feelings such as alarm, grief, and worry.[34] Batson argued that it is important to distinguish between different emotional responses (e.g., empathy and distress) because they produce different motivational states. Specifically, in response to another's suffering, empathy motivates observers to alleviate the other person's suffering (i.e., altruistic motivation). Conversely, personal distress motivates observers to alleviate their own suffering (i.e., egoistic motivation).[35]

In the course of his work, Batson found that empathy and personal distress can produce similar or distinct behavioral outcomes depending on the situation.[36] Specifically, when people cannot easily escape the perceived suffering of a victim, both empathy and personal distress may produce helping behavior for the person who is suffering. However, only empathy is likely to produce help when the observer can easily escape the perceived suffering of the victim (either by physically leaving or by mentally "exiting" the situation through one or another strategy, such as blaming the victim).

How could SCN help/have helped this debate? Batson and his colleagues conducted a study of the neural underpinnings of empathy.[37] As they discuss, their findings support the notion that human responses to others' suffering can in fact be influenced by cognitive and motivational processes. While obtaining convergent evidence from an additional method in support of prior theory and findings is quite promising and illustrative of the potential power of an SCN approach, it does not illustrate its specific strengths. We can look to SCN not only to corroborate but also to enrich our understanding of the processes involved over the time course of psychological reactions and interactions of interest. To provide compelling evidence that the differential behavior observed in the "easy exit" situation results from truly different emotional experiences and motivational consequences, it is also useful to demonstrate the similarities and differences in the neural networks underlying different ways of sharing pain with others (as did Lamm et al.).[38] Using the tools of SCN, Lamm et al. elucidated the mechanisms that allow people to distinguish self from other—a critical component for the experience of empathy. We might also consider how, if the SCN study had been conducted prior to the original studies by Batson and colleagues, it might have served as a useful catalyst for generating hypotheses that might under different conditions have been ignored. For example, if the researchers examined only situations that happened to produce similar outcomes as a result of feeling either empathy or distress, then in the absence of an SCN study, researchers less insightful than Batson and his colleagues might not have thought to look for cases in which these emotions did *not* produce similar outcomes.

Implications and Conclusions

Despite its limitations, SCN has strengths to recommend it as a vehicle for gaining insight into several psychological questions of interest. To the extent that SCN can give us insight into the actual processes being engaged in response to a particular psychological event or interaction, we may begin to understand how best to engage with, change, or counteract those processes. To date, a great deal of groundwork has been laid in brain mapping—the process by which scholars have iteratively built knowledge to determine where in the brain various processes occur (i.e., connect structure and function). If a brain region has been consistently associated with specific psychological processes that were not previously known to be involved during different kinds of tasks, the recruitment of this region in response to other stimuli or events may yield insights that we did not have before (though this view is not unchallenged).[39]

Being able to infer the psychological processes in which people are engaged is useful for several reasons. First, people may not always be willing to share honestly their emotional or cognitive states due to social desirability or other concerns. Second, people sometimes simply are unaware of the processes in which they are engaging while making certain decisions or performing certain tasks.[40] Finally, as noted above, the very act of reporting one's experience may change the nature of that experience and possibly affect later behavior.[41] SCN studies that are designed to help dissociate different underlying processes involved in the production of similar or identical behaviors, or those that help us identify similar contributing processes involved in what first appear to be different events, also may provide guiding insights. Having a tool that enables us to become aware of the processes involved when people execute various activities can be important, for example, when we are training people to improve their skills at a particular task or when we are trying to stop a process and its outcome. Indeed, as Lieberman aptly states, "In the best social cognitive neuroscience research, the *where* (in the brain) question is merely a prelude to the *when, why,* and *how* questions" (italics in original).[42]

As discussed, SCN can contribute new insight into the intent puzzle. It nonetheless is likely that the best understanding of how to decipher intent, as in the study of other complex problems, will be a multimethod approach rather than one focusing solely on SCN.[43] While no method should be viewed as a magic wand or panacea, SCN can contribute to our understanding of intent by increasing insight into the way we make attributions for others' intentions. SCN offers the promise of examining how this process might differ when we make more or less accurate assessments of intent and thus may help us understand how we might do so more effectively. With time and continued work in this area, this increased insight might extend into an improved understanding of how to measure and model others' intent and not solely our interpretation of it.

Notes

(All notes appear in shortened form. For full details, see the appropriate entry in the bibliography.)

1. Joint Publication 1-02, *Dictionary of Military and Associated Terms*, 238.
2. Ajzen and Fishbein, "Influence of Attitudes on Behavior."
3. Frith and Frith, "Interacting Minds"; and Premack and Woodruff, "Does the Chimpanzee Have a Theory of Mind?"
4. Cacioppo et al., "Just Because You're Imaging the Brain."
5. Ibid.
6. Ibid., 657.
7. Lieberman, "Social Cognitive Neuroscience," *Annual Review of Psychology*.
8. Lieberman, "Social Cognitive Neuroscience," in *Encyclopedia of Social Psychology*; and van Overwalle, "Social Cognition and the Brain," 830.
9. Lieberman, "Social Cognitive Neuroscience," in *Handbook of Social Psychology*, 173.
10. Lieberman, "Social Cognitive Neuroscience," in *Encyclopedia of Social Psychology*.
11. For example, see de Quervain et al., "Neural Basis of Altruistic Punishment."
12. King-Casas et al., "Getting to Know You."
13. Ambady and Rosenthal, "Thin Slices of Expressive Behavior."
14. Ambady and Rosenthal, "Half a Minute."
15. Ambady and Rosenthal, "Thin Slices of Expressive Behavior."
16. Premack and Woodruff, "Does the Chimpanzee Have a Theory of Mind?"
17. Ciaramidaro et al., "Intentional Network."
18. van Overwalle, "Social Cognition and the Brain."
19. Ibid.
20. Ibid.
21. Ciaramidaro et al., "Intentional Network."
22. Ekman, "Basic Emotions."
23. See, for example, Adolphs, "How Do We Know the Minds of Others?"
24. Whalen et al., "Masked Presentations of Emotional Facial Expressions"; Vuilleumier et al., "Neural Responses to Emotional Faces; and Krolak-Salmon et al., "Early Amygdala Reaction to Fear."
25. Adolphs et al., "Impaired Recognition of Emotion in Facial Expressions."
26. For example, see Heberlein et al., "Ventromedial Frontal Lobe Plays a Critical Role"; Calder et al., "Impaired Recognition of Anger"; Calder et al., "Impaired Recognition and Experience of Disgust"; Hennenlotter et al., "Neural Correlates Associated with Impaired Disgust Processing"; and Phillips et al., "Specific Neural Substrate for Perceiving Facial Expressions of Disgust."
27. Lieberman, "Social Cognitive Neuroscience," in *Handbook of Social Psychology*, 150–51.
28. Falk et al., "Neural Correlates of Persuasion."
29. Frith and Frith, "Development and Neurophysiology of Mentalizing."
30. Falk et al., "Neural Correlates of Persuasion."

31. Lieberman, "Social Cognitive Neuroscience," in *Encyclopedia of Social Psychology.*

32. Ajzen and Fishbein, "Influence of Attitudes on Behavior"; and Gollwitzer, "Implementation Intentions."

33. Batson, *Altruism Question.*

34. Batson, Early, and Salvarani, "Perspective Taking."

35. Ibid.

36. See Batson, *Altruism Question*, for a review.

37. Lamm, Batson, and Decety, "Neural Substrate of Human Empathy."

38. Ibid.

39. Lieberman, "Social Cognitive Neuroscience," in *Encyclopedia of Social Psychology*. For the challenge, see Poldrack, "Can Cognitive Processes Be Inferred?"

40. Nisbett and Wilson, "Telling More Than We Can Know."

41. Lieberman, "Social Cognitive Neuroscience," in *Encyclopedia of Social Psychology.*

42. Ibid. See also Cacioppo et al., "Just Because You're Imaging the Brain"; and Harmon-Jones and Winkielman, "Brief Overview of Social Neuroscience."

43. Campbell and Fiske, "Convergent and Discriminant Validation."

Bibliography

Adolphs, Ralph. "How Do We Know the Minds of Others? Domain-Specificity, Simulation, and Enactive Social Cognition." *Brain Research* 1079 (2006): 25–35.

Adolphs, Ralph, Daniel Tranel, Hanna Damasio, and Antonio Damasio. "Impaired Recognition of Emotion in Facial Expressions Following Bilateral Damage to the Human Amygdala." *Nature* 372 (1994): 669–72.

Ajzen, Icek, and Martin Fishbein. "The Influence of Attitudes on Behavior." In *The Handbook of Attitudes*, edited by Dolores Albarracín, Blair T. Johnson, and Mark P. Zanna, 173–221. Mahwah, NJ: Erlbaum, 2005.

Ambady, Nalini, and Robert Rosenthal. "Half a Minute: Predicting Teacher Evaluations from Thin Slices of Nonverbal Behavior and Physical Attractiveness." *Journal of Personality and Social Psychology* 64, no. 3 (1993): 431–41.

———. "Thin Slices of Expressive Behavior as Predictors of Interpersonal Consequences: A Meta-analysis." *Psychological Bulletin* 111, no. 2 (1992): 256–74.

Batson, C. Daniel. *The Altruism Question: Toward a Social Psychological Answer*. Hillsdale, NJ: Lawrence Erlbaum Associates, 1991.

Batson, C. Daniel, Shannon Early, and Giovanni Salvarani. "Perspective Taking: Imagining How Another Feels versus Imagining How You Would Feel." *Personality and Social Psychology Bulletin* 23, no. 7 (July 1997): 751–58.

Cacioppo, John T., Gary G. Berntson, Tyler S. Lorig, Catherine J. Norris, Edith Rickett, and Howard Nusbaum. "Just Because You're Imaging the Brain Doesn't Mean You Can Stop Using Your Head: A Primer and Set of First Principles." *Journal of Personality and Social Psychology* 85, no. 4 (2003): 650–61.

Calder, Andrew J., Jill Keane, Andrew D. Lawrence, and Facundo Manes. "Impaired Recognition of Anger Following Damage to the Ventral Striatum." *Brain* 127, no. 9 (September 2004): 1958–69.

Calder, Andrew J., Jill Keane, Facundo Manes, Nagui Antoun, and Andrew W. Young. "Impaired Recognition and Experience of Disgust Following Brain Injury." *Nature Neuroscience* 3 (2000): 1077–78.

Campbell, Donald T., and Donald W. Fiske. "Convergent and Discriminant Validation by the Multitrait-Multimethod Matrix." *Psychological Bulletin* 56, no. 2 (March 1959): 81–105.

Ciaramidaro, Angela, Mauro Adenzato, Ivan Enrici, Susanne Erk, Lorenzo Pia, Bruno G. Bara, and Henrik Walter. "The Intentional Network: How the Brain Reads Varieties of Intentions." *Neuropsychologia* 45 (2007): 3105–13.

de Quervain, Dominique J.-F., Urs Fischbacher, Valerie Treyer, Melanie Schellhammer, Ulrich Schnyder, Alfred Buck, and Ernst Fehr. "The Neural Basis of Altruistic Punishment." *Nature* 305, no. 5688 (2004): 1254–58.

Ekman, Paul. "Basic Emotions." In *Handbook of Cognition and Emotion*, edited by Tim Dalgleish and Mick J. Power, 45–60. San Francisco, CA: John Wiley & Sons Ltd, 1999.

Falk, Emily B., Lian Rameson, Elliot T. Berkman, Betty Liao, Yoona Kang, Tristen K. Inagaki, and Matthew D. Lieberman. "The Neural Correlates of Persuasion: A Common Network across Cultures and Media." *Journal of Cognitive Neuroscience*, early access posted online 19 November 2009. http://www.mitpressjournals.org/doi/abs/10.1162/jocn.2009.21363.

115

Frith, Chris D., and Uta Frith. "Development and Neurophysiology of Mentalizing." *Philosophical Transactions of the Royal Society of London*, Series B, Biological Sciences 358, no. 1431 (29 March 2003): 459–73.

———. "Interacting Minds—A Biological Basis." *Science* 286, no. 5445 (1999): 1692–95.

Gollwitzer, Peter M. "Implementation Intentions: Strong Effects of Simple Plans." *American Psychologist* 54, no. 7 (July 1999): 493–503.

Harmon-Jones, Eddie, and Piotr Winkielman. "A Brief Overview of Social Neuroscience." In *Social Neuroscience: Integrating Biological and Psychological Explanations of Social Behavior*, edited by Eddie Harmon-Jones and Piotr Winkielman, 3–11. New York: Guilford Press, 2007.

Heberlein, Andrea S., Alisa A. Padon, Seth J. Gillihan, Martha J. Farah, and Lesley K. Fellows. "Ventromedial Frontal Lobe Plays a Critical Role in Facial Emotion Recognition." *Journal of Cognitive Neuroscience* 20, no. 4 (2008): 721–33.

Hennenlotter, Andreas, Ulrike Schroeder, Peter Erhard, Bernhard Haslinger, Robert Stahl, Adolf Weindl, H. Grafin von Einsiedel, Klaus W. Lange, and Andres O. Ceballos-Baumann. "Neural Correlates Associated with Impaired Disgust Processing in Presymptomatic Huntington's Disease." *Brain* 127, no. 6 (2004): 1446–53.

Johnson-Frey, Scott H., Farah R. Maloof, Roger Newman-Norlund, Chloe Farrer, Souheil Inati, and Scott T. Grafton. "Actions or Hand-Object Interactions? Human Inferior Frontal Cortex and Action Observation." *Neuron* 39, no. 6 (2003): 1053–58.

Joint Publication 1-02. *Department of Defense Dictionary of Military and Associated Terms*, 12 April 2001, as amended through 31 October 2009.

King-Casas, Brooks, Damon Tomlin, Cedric Anen, Colin F. Camerer, Steven R. Quartz, and P. Read Montague. "Getting to Know You: Reputation and Trust in a Two-Person Economic Exchange." *Science* 308, no. 5718 (1 April 2005): 78–83.

Krolak-Salmon, Pierre, Marie-Anne Hénaff, Alain Vighetto, Olivier Bertrand, and François Mauguière. "Early Amygdala Reaction to Fear Spreading in Occipital, Temporal, and Frontal Cortex: A Depth Electrode ERP Study in Human." *Neuron* 42, no. 4 (2004): 665–76.

Lamm, Claus, C. Daniel Batson, and Jean Decety. "The Neural Substrate of Human Empathy: Effects of Perspective-Taking and Cognitive Appraisal." *Journal of Cognitive Neuroscience* 19, no. 1 (January 2007): 42–58.

Lieberman, Matthew D. "Social Cognitive Neuroscience." In *Encyclopedia of Social Psychology*, edited by Roy F. Baumeister and Kathleen D. Vohs. Thousand Oaks, CA: Sage Press, 2007.

———. "Social Cognitive Neuroscience." In *Handbook of Social Psychology*, 5th ed., edited by Daniel T. Gilbert, Susan T. Fiske, and Gardner Lindzey. New York: McGraw-Hill, 2010.

———. "Social Cognitive Neuroscience: A Review of Core Processes." *Annual Review of Psychology* 58 (2007): 259–89.

Nisbett, Richard E., and Timothy D. Wilson. "Telling More Than We Can Know: Verbal Reports on Mental Processes." *Psychological Review* 84, no. 3 (May 1977): 231–59.

Phillips, Mary L., Andrew W. Young, Carl Senior, Michael J. Brammer, Christopher Andrew, Andrew J. Calder, Edward T. Bullmore, et al. "A Specific Neural Substrate for Perceiving Facial Expressions of Disgust." *Nature* 389, no. 6650 (1997): 495–98.

Poldrack, Russell A. "Can Cognitive Processes Be Inferred from Neuroimaging Data?" *Trends in Cognitive Sciences* 10, no. 2 (February 2006): 59–63.

Premack, David, and Guy Woodruff. "Does the Chimpanzee Have a Theory of Mind?" *Behavioral and Brain Sciences* 1 (1978): 515–26.

van Overwalle, Frank. "Social Cognition and the Brain: A Meta-Analysis." *Human Brain Mapping* 30 (2009): 829–58.

Vuilleumier, Patrik, Jorge L. Armony, Karen Clarke, Masud Husain, Julia Driver, and Raymond J. Dolan. "Neural Responses to Emotional Faces with and without Awareness: Event-Related fMRI in a Parietal Patient with Visual Extinction and Spatial Neglect." *Neuropsychologia* 40, no. 12 (2002): 2156–66.

Whalen, Paul J., Scott L. Rauch, Nancy L. Etcoff, Sean C. McInerney, Michael B. Lee, and Michael A. Jenike. "Masked Presentations of Emotional Facial Expressions Modulate Amygdala Activity without Explicit Knowledge." *Journal of Neuroscience* 18, no. 1 (1998): 411–18.

Chapter 9

Neuropsychological and Brain-Imaging Techniques

Abigail J. C. Chapman, NSI, Inc.

Abstract: The objective of this paper is to provide an overview of the current neuropsychological tools and techniques available to social scientists. These tools, such as
lesion studies and functional magnetic resonance imaging
(fMRI), help researchers understand the complex relationship between mind and body to explore the possibility of
harnessing the ability to anticipate an individual's intention to engage in a behavior. While the field of social cognitive neuroscience offers exciting research findings through
the use of promising tools and techniques, it is a nascent
field that requires further study and experimentation before conclusive findings can be applied to current problems facing the defense community.

James Watson once said, "There are only molecules; everything
else is sociology." His tongue-in-cheek arrogance reminds us of
the great gulf that once separated the "two cultures" of humani-
ties and science. In the last decade this gap is successfully be-
ing bridged by social neuroscience.

—V. S. Ramachandran, MD, PhD
University of California, San Diego

Introduction

Researchers have long been fascinated by the complexities of
the human brain and, in particular, the relationship between
mind and body and the possibility of harnessing the ability to
predict an individual's intention to engage in a behavior. While
it has been and will remain for many years to come an organ

that cannot be easily seen or understood, science has made great strides in developing tools and techniques that allow us to gain insight into the neural mechanisms (the patterns of brain activation) underlying social cognitive processes,[1] such as social perception,[2] attitude formation,[3] emotion recognition,[4] and decision making,[5] as well as the theory of the mind.[6]

Only in the last 10 years have researchers begun marrying the tools of social psychology and neuroscience to create a new field of study that harnesses the strengths of both to examine phenomena, such as intention, that cannot be easily addressed by studying behavior (e.g., obedience studies) or relying upon self-reported responses (e.g., surveys, implicit attitude tests). While applying the tools of social psychology and neuroscience allows researchers to gain deeper insight by capitalizing on the multimethod approach to studying various phenomena, it is important to note that at this relatively nascent stage, studies are restricted to a laboratory setting, are reliant upon volunteers, and incur high monetary costs associated with certain neuroimaging tools (to be explained in detail below). In addition, to discover and isolate the neural imprint, or neural correlates, of an attitude, behavior, or intention to engage in a behavior, it is necessary to conduct rigorous experiments in a controlled setting across numerous studies to ensure the findings are consistent.

This chapter will provide an overview of the current neuropsychological tools and techniques available to social scientists. The tools are presented in detail below starting with the oldest technique (lesion studies) and ending with the most recent and most often used technique (functional magnetic resonance imaging [fMRI]).

Techniques

Lesion studies

Lesion studies are correlation studies that are primarily conducted on individuals who suffer from an isolated and identifiable brain injury from an accident, disease, or surgical lesion. This body of work laid the foundation for the field of social cognitive neuroscience and allows for the close examination of the correlation between damage to isolated areas of the brain and observable changes in psychological functioning (i.e., changes in emotion, mood, and cognition).

Procedure. In some cases, to study changes in performance on tasks of interest, a temporary lesion is created using transcranial magnetic stimulation (TMS), a technique that uses electromagnetic pulses to stimulate the neurons in a targeted area of the cerebral cortex to simulate a lesion (the cortex is the area of the brain that is also referred to as the grey matter and is responsible for higher-order or executive brain functions). In situations where the exact location of the lesion is not known due to injury through an accident (e.g., a foreign object forcibly inserted through the skull) or disease (e.g., tumor), it is necessary to obtain a structural brain scan of the individual so that the damaged area of the brain can be identified.

The most well-known example of a lesion study is the posthumous case study of Phineas Gage, a railroad worker in 1848. While working on the California railroad, Gage survived the impalement of a railroad rod through his head, but the injury severely damaged both his left and right frontal lobes, resulting in an observable change in demeanor.[7] An 1868 report by Gage's physician in fact noted that Gage's "mind was radically changed, so decidedly that his friends and acquaintances said he was 'no longer Gage.'"[8] However, it was the lesion study conducted by Damasio et al. in 1994 that provided scientific evidence supporting the historical account that Mr. Gage's behaviors and actions were radically altered by the accident. Damasio et al. compared computer-generated images of Gage's skull to those of other patients who had suffered damage in the same area of the brain and in fact had exhibited similar symptoms. Thus, it was through the observation of changes in behavior and the comparison of identical brain damage in other patients that researchers were able to identify the critical role that the frontal lobe plays in decision making, emotion regulation, and behavior.

Benefits and Limitations. Primarily, a lesion-based study is used to establish a causal relationship between damage to a localized area of the cortex and demonstrated changes in the personality, behavior, and executive processing tasks of an individual. Given that the study is done on an individual level, the research is reliant upon finding numerous individuals who all suffer from isolated damage to the identical location of their brain or using the TMS procedure to induce a temporary "lesion" in a sample group. A limitation to conducting this type of study is that research is restricted to examining higher-order processing tasks and is easily confounded if the damage is not localized. An appealing benefit to a lesion-based study is that it is relatively low cost and aids in establishing a causal relationship, as long as the findings are replicated across numerous studies.

Electroencephalography

Using an electroencephalograph (EEG) machine, researchers measure the electrical activity of neurons, or brain waves, (known as event-related potentials) that are activated in the outer layers of the brain when the participant is presented with a task.

Procedure. The EEG machine requires that the participant wear a skull-cap that contains a number of electrodes (from 20 to 256 or more) soaked in a conductive gel solution allowing the electrode to connect easily with the skin. Prior to the start of the study, the experimenter inputs the time stamps of the stimuli into the computer protocol so that the data can be "tagged" as either occurring in response to a stimulus or resting background activity of the individual. Once the data has been averaged across several trials, it is relatively easy to identify patterns of brain-wave activity as reliable responses to the given stimuli. The pattern of wave activation is then mapped onto a generic computer-generated image of the brain that allows for the illustration of the degree of activation, or density of response, in the brain as it responds to the stimulus.[9]

Benefits and Limitations. The EEG machine is a relatively low-cost, noninvasive technology that allows for the mapping of real-time neural responses to a given stimulus (the temporal resolution is about one millisecond). There are two main drawbacks to conducting this type of study. First, only the outer layer of the cerebral cortex is measurable, unless electrodes are implanted subdermally, which limits the research to questions that involve only the outer layer of the cerebral cortex (i.e., executive functioning tasks). Second, spatial resolution of event-related potentials (the measured brain activity) is poor, thus not allowing the exact location of the activity to be identified. The spatial resolution can be improved by increasing the number of electrodes attached to the skullcap, but it still does not allow for the exact pinpointing that the following technique provides.

Functional Magnetic Resonance Imaging

Functional magnetic resonance imaging (fMRI) has been used successfully to identify, diagnose, and understand neurological conditions and brain abnormalities, but only in the last few years have social cognitive psychologists employed fMRI to detect subtle changes in localized activity within the brain while participants engage in an experiment designed to elicit a response in a particular area of the brain that researchers believe may correspond to the emotion, behavior, or attitude they are studying. The fMRI relies upon a model of homodynamic response that posits that a "transient increase in neuronal activity within a region of the brain begins consuming additional oxygen in the blood proximal to these cells. . . . As a result, blood near a region of local neuronal activity soon has a higher concentration of oxygenated hemoglobin than blood in locally inactive areas."[10] In other words, just like any other muscle in the body, when an area of the brain is activated or engaged, it receives an influx of oxygenated blood. Further, the model suggests that the more oxygenated blood has a different magnetic property, allowing the fMRI machine to capture visual images of the change in oxygenated blood at a one- to two-second latency. The brain is scanned on a preset interval schedule that corresponds to the experimental protocol, allowing for control scans when the brain is at rest and experimental scans when the participant is engaged in a cognitive task. Results are obtained by subtracting the control, or resting, scan from the experimental scan, leaving behind the difference in activation. We can theorize that the activation differential is primarily due to the task that the participant was engaged in.

Procedure. An fMRI scan requires a participant's head or entire body to be placed inside an fMRI scanner, which is built around a powerful electrical magnet (approximately 10^5

times stronger than Earth's magnetic field), for an extended amount of time.[11] The entire scan usually lasts 30–45 minutes and requires the participant to remain immobile to ensure that the images obtained are not affected by movement. It is important to ensure that the participants are not claustrophobic and fully understand the task that will be asked of them while they are inside the scanner. It is helpful to do a practice set outside the scanner on a computer where the researcher is available to clarify and answer potential questions. Once the participant is inside the scanning machine, it is important to first complete a structural brain scan of the individual to ensure there are no abnormalities in structure that would preclude the individual from participating in the study. This usually takes less than five minutes. In addition, due to the uniqueness of every brain (e.g., areas of the brain are not uniform in size or exact location), the structural scan is used to map the areas of activation onto the individual's brain to ensure consistency in the identification and labeling across participants.

Once the experiment has begun, the participant is presented with a slide designed to elicit minimal neural response, such as a crosshair (fig. 9.1), so that a baseline neural response can be calculated and contrasted with the experimental condition. In the simple example presented in figure 9.1, the participant first views a baseline image for three seconds, an image of a square for five seconds, and the baseline image for three seconds, and then is asked to engage his or her recall, or short-term memory, in identifying the object that he or she has just seen. This series is repeated several times with different experimental slides (e.g., different geometric shapes) over the course of approximately 30 to 60 minutes to ensure that there will be enough clean scans of the brain.

After all participants are run through the experimental protocol, the data is scrubbed and prepared for analysis, a process called preprocessing. It is within this step that the data is made usable for the final analysis. Preprocessing of the experimental data includes realignment (correcting the image to account for motion of the individual within the scanner), normalization (alignment of each individual's brain onto one common brain to line up brain regions and structures across the sample group), and smoothing (averaging) of the data to ensure that the data

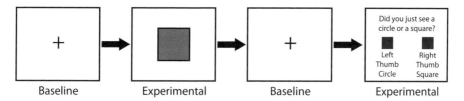

| Baseline | Experimental | Baseline | Experimental |

Figure 9.1. An example of an fMRI scanning protocol. The participant sees each baseline slide for three seconds and each experimental slide for five seconds. Only during the second experimental slide is the participant asked to engage in a response behavior—using his or her thumbs to depress the corresponding button on the button-box device by his or her side.

can be compared across individuals. It is important to point out that the three steps involved in preprocessing change the data to make it more analyzable. As Lieberman points out, "[The data] are far from their raw state and represent a series of decisions and transformations that render the data more analyzable but sometimes introduce problems when the data do not conform to the assumptions behind the transformations."[12]

Finally, once the data has been analyzed, the results are either overlaid on a standard structural brain scan, a composite brain scan of all the participants, or a specific individual's structural brain scan. The images are usually presented in color, with the areas of activation displayed in a red or "hot" color and areas of decreased activation in shades of blue.

Benefits and Limitations. The primary benefit of conducting an fMRI study is that the technique allows researchers to look inside the brain to see not only which brain regions respond to certain stimuli but also the patterns and strengths of activation. It allows for three-dimensional imaging of subcortical activation in response to stimuli, providing greater understanding of the intricacies of the brain as it responds to the social environment. While fMRI studies have significant appeal and benefits both scientifically and visually, there are also several major limitations associated with this technology, including the high cost of acquiring and maintaining the scanner.

Since it is necessary to obtain multiple data points to average across the responses, each experimental condition must be presented to the participants across multiple trial repetitions, which has the potential for contamination and habituation effects. Ensuring that all participants are cognitively engaged in the study, that each experimental condition is consistent, and that the total run time for the scan is under one hour can help mitigate this drawback.

Because of the limitations of current fMRI technology, behavioral research that utilizes the technology is constrained by operational requirements. For instance, it is difficult to study an individual's intention to engage in a specific behavior and even more challenging while he or she is alone inside an fMRI machine and exposed to visual and auditory stimuli via goggles and headphones. The scanner itself creates a unique environment that does not reflect a real-life situation. Consequently, researchers are limited in their ability to generalize their findings beyond the laboratory setting and are restricted by their need to rely upon volunteers to participate in studies.

In addition, the current technology is cumbersome and noisy, unsuited to clandestine scanning applications. Therefore, the technique may be applied only in scenarios where the participant's awareness of the scanning process is of no concern. In other words, current fMRI technology cannot be used to remotely scan individuals in a public setting such as an airport to look for intention to engage in threatening behavior. Consequently, given the controlled and sterile environment of an fMRI laboratory, researchers studying social behavior have difficulty generalizing their findings beyond the laboratory setting and examining intentions to engage in behavior outside of the fMRI.

One way to compensate for this limitation is to engage in a two-part study comprised of an fMRI study followed by a survey over a short period of time. In the first part of the study, researchers would identify a behavior they are interested in studying. For example, researchers may be interested in exploring the link between pro-environmental attitudes and past behavior and an individual's intention and subsequent use of compact fluorescent light (CFL) bulbs. The researchers would run participants through an fMRI protocol designed to elicit a neural response to questions on their environmental attitudes, past behaviors, and possible intention to use CFL bulbs in the future. Several weeks later, researchers would contact each participant and ask them several follow-up behavioral questions (such as, Have you purchased CFL bulbs for your home?) to see if their behaviors mapped onto their stated responses and neural activity from the fMRI session. Alternatively, for the second part of the study, the researchers could monitor a basket of free CFL bulbs sitting by the door and record how many each participant took and follow up with them to see if they used them in their homes.

Conclusion

While the field of social cognitive neuroscience offers exciting research findings through the use of promising tools and techniques, readers of the published results of these neuroimaging studies should maintain a critical eye. Though the tools described above provide greater insight into how the brain works, it is important to realize that the brain is subject to change, structurally and neurally, by the environment and experience, both in early development and throughout adulthood (this is referred to as neural plasticity). In addition, while the current techniques used within the field of social cognitive neuroscience allow for the study of phenomena that cannot be easily observed, the tools themselves are relatively cumbersome and cannot be easily taken outside of a controlled laboratory setting. With current technology, it is nearly impossible to conduct experiments outside of a laboratory setting or to scan individuals covertly. Given the current limitations associated with neuroimaging studies and the relatively limited body of research, it is critical that we continue to conduct experiments using both neuroimaging and traditional social psychology techniques. We must ensure that the results are appropriately caveated and that researchers continue to replicate study findings before we can lay claim to understanding the way the brain works.

Notes

(All notes appear in shortened form. For full details, see the appropriate entry in the bibliography.)

1. Lieberman, "Social Cognitive Neuroscience," *Annual Review of Psychology*.
2. Bartholow and Dickter, "Social Cognitive Neuroscience of Person Perception"; and Ito, Willadsen-Jensen, and Correll, "Social Neuroscience and Social Perception."
3. Cunningham and Johnson, "Attitudes and Evaluation."

4. Morris et al., "Differential Neural Response"; and Nomura et al., "Functional Association of the Amygdala and Ventral Prefrontal Cortex."

5. Krawczyk, "Contributions of the Prefrontal Cortex."

6. Stone, "Evolutionary Perspective on Domain Specificity"; and Van Overwalle, "Social Cognition and the Brain."

7. Damasio et al., "Return of Phineas Gage."

8. Harlow, "Recovery from the Passage of an Iron Bar through the Head," 277.

9. Dubin, *How the Brain Works.*

10. Cacioppo et al., "Just Because You're Imaging the Brain," 651.

11. Langleben, "Detection of Deception with fMRI," 2.

12. Lieberman, "Social Cognitive Neuroscience," in *Handbook of Social Psychology*, 147.

Bibliography

Bartholow, Bruce D., and Cheryl L. Dickter. "Social Cognitive Neuroscience of Person Perception: A Selective Review Focused on the Event-Related Brain Potential." In *Social Neuroscience: Integrating Biological and Psychological Explanations of Social Behavior*, edited by Eddie Harmon-Jones and Piotr Winkielman, 376–400. New York: Guilford Press, 2007.

Cacioppo, John T., Gary G. Berntson, Tyler S. Lorig, Catherine J. Norris, Edith Rickett, and Howard Nusbaum. "Just Because You're Imaging the Brain Doesn't Mean You Can Stop Using Your Head: A Primer and Set of First Principles." *Journal of Personality and Social Psychology* 85, no. 4 (2003): 650–61.

Cunningham, William A., and Marcia K. Johnson. "Attitudes and Evaluation: Toward a Component Process Framework." In *Social Neuroscience: Integrating Biological and Psychological Explanations of Social Behavior*, edited by Eddie Harmon-Jones and Piotr Winkielman, 227–45. New York: Guilford Press, 2007.

Damasio, Hanna, Thomas Grabowski, Randall Frank, Albert M. Galaburda, and Antonio R. Damasio. "The Return of Phineas Gage: Clues about the Brain from the Skull of a Famous Patient." *Science* 264, no. 5162 (20 May 1994): 1102–5.

Dubin, Mark W. *How the Brain Works.* Williston, VT: Blackwell Publishing, 2002.

Harlow, John M. "Recovery from the Passage of an Iron Bar through the Head." *Publications of the Massachusetts Medical Society* 2 (1868): 327–47.

Ito, Tiffany A., Eve Willadsen-Jensen, and Joshua Correll. "Social Neuroscience and Social Perception: New Perspectives on Categorization, Prejudice and Stereotyping." In *Social Neuroscience: Integrating Biological and Psychological Explanations of Social Behavior*, edited by Eddie Harmon-Jones and Piotr Winkielman, 400–24. New York: Guilford Press, 2007.

Krawczyk, Daniel C. "Contributions of the Prefrontal Cortex to the Neural Basis of Human Decision-Making." *Neuroscience and Biobehavioral Reviews* 26, no. 6 (2002): 631–64.

Langleben, Daniel D. "Detection of Deception with fMRI: Are We There Yet?" *Legal and Criminological Psychology* 13, no. 1 (2008): 1–9.

Lieberman, Matthew. "Social Cognitive Neuroscience." In *Handbook of Social Psychology*, 5th ed., edited by Daniel T. Gilbert, Susan T. Fiske, and Gardner Lindzey, 143–93. Vol 1. New York: McGraw-Hill, 2010.

———. "Social Cognitive Neuroscience: A Review of Core Processes." *Annual Review of Psychology* 58 (2007): 259–89.

Morris, Stephen, Chris D. Frith, David I. Perrett, Daniel Rowland, Andrew W. Young, Andrew J. Calder, and Raymond J. Dolan. "A Differential Neural Response in the Human Amygdala to Fearful and Happy Facial Expressions." *Nature* 383 (1996): 812–15.

Nomura, Michio, Hideki Ohira, Kaoruko Haneda, Tetsuya Iidaka, Norihiro Sadato, Tomohisa Okada, and Yoshiharu Yonekura. "Functional Association of the Amygdala and Ventral Prefrontal Cortex during Cognitive Evaluation of Facial Expressions Primed by Masked Angry Faces: An Event-Related fMRI Study." *NeuroImage* 21, no. 1 (January 2004): 352–63.

Stone, Valerie E. "An Evolutionary Perspective on Domain Specificity in Social Intelligence." In *Social Neuroscience: Integrating Biological and Psychological Explanations of Social Behavior*, edited by Eddie Harmon-Jones and Piotr Winkielman, 316–49. New York: Guilford Press, 2007.

Van Overwalle, Frank. "Social Cognition and the Brain: A Meta-Analysis." *Human Brain Mapping* 30 (2009): 829–58.

Chapter 10

Desperate Measures

Different Types of Violence, Motivations, and Impact on Stability

Thomas Rieger, Gallup Consulting*

Abstract: All types of violence are not created equal, and a decline in violence is not necessarily equivalent to an increase in stability. Violence due to instability and violence that stems from the pursuit of some ideological goal are often driven by very different factors and have different purposes. Instability due to the lack of ability to survive and thrive may lead to substantial unrest, violence, rioting, or other undesirable activity. Iraq provides an example to explain extremist violence and to illustrate how it is possible to win a war and lose the peace. The example suggests that counterterrorism efforts must be undertaken in coordination with stability efforts to effectively manage both types of potential unrest.

Introduction

Desperate times call for desperate measures. It is a cliché often invoked to justify some strong action in the face of adversity. There are numerous contemporary examples of individuals and groups taking desperate measures, from food riots in Haiti to suicide bombings in Baghdad. While both these examples represent violent actions in pursuit of a goal, they are quite different in many ways.

Violence due to instability and violence that stems from the pursuit of some ideological goal are often driven by very different factors with very different purposes. It may be tempting,

---

*The author would like to thank Dr. Shane J. Lopez, PhD, and Lt Col Christopher Rate, USAF, PhD, for their valuable contributions to this work.

131

however, to overlook those distinctions in interpreting changes in levels of violence. For example, one might assume that a decline in extremist violence such as improvised explosive device (IED) attacks would represent increased stability. That assumption in many cases would be incorrect.

Not all acts of violence are born out of the same conditions or inspired by the same psychological motivations. Some are driven more by a need to survive, while others are in pursuit of a larger purpose. Both may require some level of courage before an individual can convince himself or herself to commit the act, but there are multiple types of courage that can shed light on the distinctions among different types of violent actions.

Types of Courage

The first step in understanding the differences between violence due to difficult living conditions and violence due to extremism is to understand the psychological factors that produce the courage to commit such an act. Defining courage is no simple task. The long history of discussion on the topic dates all the way back to Aristotle (350 BC).[1] For the purposes of this essay, however, we will draw upon recent research that provides relevant definitions for understanding different types of violence.

C. R. Snyder and Shane Lopez, in their definitive work on positive psychology, define several types of courage, including *vital courage* and *moral courage.* Vital courage is defined as the inspiration for an action taken that "extends lives." In other words, the ultimate and primary motivation for acts of vital courage is to preserve survival for oneself or others to which an individual feels connected and responsible. By definition, the purpose of vital courage is not to sacrifice a life. It is to save a life.[2]

Moral courage, on the other hand, is defined as an act that "preserves the ideals of perceived justice and fairness." Snyder and Lopez further define moral courage as "the authentic expression of one's beliefs or values in pursuit of justice or the common good despite power differentials, dissent, disapproval, or rejection." In other words, while vital courage is inwardly focused (survival), moral courage is outwardly focused (ideology).[3]

It can be argued that *courage* is the wrong label for the actions of a suicide bomber or someone stealing water from a

132

reservoir, as these are far from desirable behaviors. We do not ascribe worth to or express admiration for these activities. These definitions of courage are put forth simply to provide a framework for understanding some of the very fundamental differences among different types of violence, each of which represents a course of action that may bring harm or even death to the actor.

The label *courage* is also appropriate based on established criteria for what can be considered a courageous act. Violence due to extremism and violence in pursuit of survival include willing, intentional acts involving substantial danger, difficulty, or risk to the actor that are primarily motivated to bring about a noble, good, or morally worthy purpose, as defined by the actor—all marks of a courageous act.[4] Given the commonality on these three dimensions between courageous and violent acts, courage appears to be a reasonable label for part of the psychological motivation behind violent acts in support of extremist agendas and violent acts intended for survival.

Neither vital nor moral courage should be associated solely with heroes or villains. Both are motivating forces that could apply to either type of person.

Violence Inspired Primarily by Vital Courage

During 2008 the cost of basic food staples rose sharply in many parts of the world.[5] The desperate violence that ensued provides a clear example of actions taken by those who seek to extend lives, a key component of the definition of vital courage.

The shortages and higher prices inspired riots in many countries, the most severe of which took place in Bangladesh, Haiti, India, Indonesia, Mozambique, Senegal, Somalia, Yemen, Ivory Coast, Cameroon, Mexico, Burkina Faso, Mauritania, and Egypt. In these riots, many lost their lives, were injured, or were imprisoned. There was also substantial damage to public and private property. The rioting and associated violence were not inspired by the desire to recreate an Islamic caliphate or in pursuit of some holy jihad. Instead the violence was ultimately about food.

Certainly, in many of these cases, someone or something needed to provide the spark to motivate the crowd to take action. Claims of unfairness, objections to a change in taxation or

133

policy, assertions of one's right to survive, and so forth were undoubtedly made to justify the actions taken. But while ideology of a sort played a role, the primary motivating force was a desire to survive or improve one's quality of life. Hence, vital courage (as we are defining it here) was likely the stronger force.

As an example, in 2008 violence broke out in the mining regions of Peru, where as many as 20,000 protesters closed roads, kidnapped and beat law enforcement personnel, and for all intents and purposes, closed down the entire region due to perceived inequities in revenue sharing.[6] Once again, religion or ideology had virtually nothing to do with the root cause of the violence even though taxation and revenue-sharing policy provided the catalyst.

Violence Inspired Primarily by Moral Courage

While there are many examples of acts of violence spurred by a desire both to survive and to achieve an ideological purpose (for example, guerilla activity against an invading army), some cases by definition put the desire to survive aside to achieve some ideological purpose or have nothing whatsoever to do with survival. The clearest example of an act that has nothing to do with survival is a suicide bombing, where the actor knows full well that he or she will die. Once someone volunteers, elements of vital courage may be injected into the situation by the leaders of the extremist group (if the individual backs out, his or her family will be killed), but the primary motivation for volunteering is based more in ideology than the ability to survive, violating the "desire to extend lives" aspect of vital courage.

As another example, the nerve gas attack perpetrated by Aum Shinrikyo in Tokyo clearly had nothing to do with the ability to get food, shelter, or water. It was purely inspired by the "expression of authentic beliefs," as our definition of moral courage dictates. The same could be said for the Oklahoma City bombing. Timothy McVeigh did not plan and carry out the bombing so that he would have more food and water or more comfortable living conditions.

In fact, conducting acts that are ideologically driven often would be contrary to behavior driven by vital courage. Once individuals join a terrorist group and participate in civilian attacks, they do so knowing full well that the combined might of national and international police, military, and intelligence forces will likely hunt them down relentlessly for the rest of their lives.

Terrorist acts of this type are meant to achieve an ideological purpose. Those that commit these acts, therefore, are doing so to further progress in achieving that purpose. Snyder and Lopez's definition of moral courage (the authentic expression of one's beliefs or values in pursuit of justice or the common good despite power differentials, dissent, disapproval, or rejection) fits these types of acts very well. The acts are an expression of the actor's beliefs in pursuit of some goal he or she believes to be worthy, despite a power differential and the inevitable consequences for the actor.

According to the National Counterterrorism Center, the countries with the most deaths due to terrorist acts in 2007 included Iraq, Afghanistan, Pakistan, India, Thailand, Somalia, Sudan, Chad, Colombia, Sri Lanka, Philippines, Algeria, Democratic Republic of Congo, and Russia.[7] With the exception of India and Somalia, none of these countries were also among those that experienced the worst violence due to poor living conditions or absence of basic needs (listed above). The same could be said for the countries in which the most kidnappings took place (Iraq, Nepal, Philippines, Gaza Strip, Afghanistan, Democratic Republic of Congo, Pakistan, India, Chad, Nigeria, Colombia, Niger, Sudan, Netherlands, and Kenya).[8] In each of these countries, the vast majority of violence was due primarily to the desire to further some ideology or larger purpose.

There has been much in the news recently about the role that Islamic extremism has played for many in providing a "greater purpose" to inspire acts framed under the banner of moral courage. However, it should not be assumed that terrorist activity inspired by moral courage is possible only as an extremist interpretation of Islam. History abounds with examples of ideologically driven attacks on civilians that had nothing whatsoever to do with Islamic doctrine, such as the Shining Path in Peru, the Revolutionary Armed Forces of Colombia, or even the Ku Klux Klan in the United States. In each of these cases, justification for a civilian attack was provided through the need to further some purpose that was important to the actor.

One potential flaw in this line of reasoning is that society provides a different definition of what is right and wrong, and if someone is primarily driven by perceptions of ideology, societal pressure would prevent him or her from ultimately committing the act.

The ability of the nonstate actor to overcome this apparent disconnect can be partially explained by behavioral economist George Loewenstein, who found that "whenever individuals face tradeoffs between what is best for themselves and what is morally correct, their perceptions of moral correctness are likely to be biased in the direction of what is best for themselves."[9]

Further insight into this dynamic can be gained from Lawrence Kohlberg's work on the stages of moral development. If someone has not advanced beyond what Kohlberg describes as stage two, or "self-interest driven" morals, he or she would not necessarily be very concerned about achieving social order or higher ethical principles (the more advanced stages).[10]

One of the main reasons why individuals subscribe to a particular ideology is that they believe it is a worthy purpose and that somehow their lives, as well as the lives of others like them, will be better as a result. Therefore, it is not that far of a leap to allow the end to justify the means, as Loewenstein suggests might happen in the face of a trade-off between self-interest and moral correctness. The extremist interpretation of what is allowable under the Koran is an often-cited source for overcoming this trade-off among jihadist groups.[11]

Certainly the desire to create a better world contains aspects of vital courage, but the primary motivation is not survival. It is making progress toward an ideological (and presumably desirable) goal. Given the above discussion, it is reasonable to assume that acts meant to achieve some larger purpose or ideologically based outcome are driven more by moral courage than vital courage—in fact, committing such an act is often in direct conflict with one's ability to survive and thrive (vital courage).

Example: Iraq

According to the 2008 9010 Report to Congress, since the surge (and with the help of the Awakening Councils, local groups of Sunnis) there have been dramatic declines in the

136

number of attacks and casualties resulting from terrorist or extremist activity. The report also notes that provisions for basic needs such as fresh water and electricity are still lacking in most parts of the country. In addition, a sizable proportion of the country still must rely on food rations to survive.[12]

In other words, violence inspired by moral courage has declined. But the potential still exists for it to be replaced by violent acts driven by vital courage. To illustrate this point, one must merely imagine what would happen if food rations simply stopped. It would not take very long for violence to break out.

Given the lack of basic services, cholera outbreaks from poor sewage systems, and the inherent instability that these conditions may cause, assuming that the decline in terrorist violence is equal to a decline in the potential for *any* violence would be a potentially disastrous conclusion. Specifically, if a society is not able to sustain itself, it is by definition unstable. Instability due to the lack of ability to survive and thrive may lead to substantial unrest, violence, rioting, or other undesirable activity.

A Vicious Cycle

It is, of course, very possible to win a war and lose the peace. If the assumptions discussed in the example prove correct, societies such as Iraq that experienced declines in extremist violence may be very prone to future violence due to unstable conditions, which in turn may lead to more violence inspired by ideology. According to the Gallup Political Radicals (POLRAD) model, Type Two radicals are downscale and perceive themselves to be victims. As this group is highly leader seeking, it may provide a very rich recruiting pool for the more mainstream Type One radicals, who are more ideologically driven.[13] Separate research by Troy Thomas and Stephen Kiser describes this same phenomenon as "ideology entrepreneurs" who seek to win over the victimized poor.[14]

In either case, creating downscale victims risks sowing the seeds of a resurgence of terrorist group membership and activity by not addressing the survival needs of the population at risk. The resurgence of the Taliban in Afghanistan is perhaps the most recent example of this dynamic. If a leader of an ex-

137

tremist group can provide promises of a better world (and inspire moral courage) while blaming others for poor living conditions (inspiring vital courage), he or she would likely be sending a very compelling message to a desperate population. If a postwar state cannot win the peace, the war may start all over again.

Implications for Policy

The points presented in this paper contain several implications for policy. First, it is dangerous to assume that all types of violence are created equal and that a decline in terrorist violence is equivalent to increased stability. The converse is also true. A decline in unrest due to poor conditions may be unrelated to violence inspired by ideology.

Violence due to extremism and violence due to instability represent similar behaviors stemming from very different psychological motivations and environmental conditions. Eliminating one does not necessarily mean that a society is safe from the other. Achieving a lasting peace may require elimination or mitigation of extremist elements, while at the same time building a self-sustaining society that can take care of itself. By not viewing safety as a multidimensional issue, policy makers risk declaring victory when there may be other battles to be fought.

Second, and related to the first implication, is the necessity to tie together counterterrorism efforts and stability operations. Eliminating extremist elements from a population that is otherwise highly unstable runs the risk of replacing suicide bombings with food riots, increased crime, and possible insurgency. Increasing stability in an environment that is experiencing acts of desperation without addressing ideologically based extremist elements will likely see the opposite occur.

Even if extremist elements are eliminated, an unstable environment may over time provide a swamp from which future nonstate actors are born (Type Two radicals, as described above). On the other hand, if a society can provide a sustaining economy, basic services, and decent living conditions to its population while minimizing or eliminating extremist violence, then it is much more likely to achieve a lasting peace.

Notes

1. Aristotle, *Nicomachean Ethics*, book 3, chapters 7–9.
2. Snyder and Lopez, *Positive Psychology*, 241.
3. Ibid., 242.
4. Rate, "What Is Courage?" 77.
5. Walt, "World's Growing Food-Price Crisis."
6. Palomino and Aquino, "Peru Police Held Hostage."
7. National Counterterrorism Center, *2007 Report*, 26.
8. Ibid., 29.
9. Babcock and Loewenstein, "Explaining the Bargaining Impasse," 230. See Loewenstein's study "Behavioral Decision Theory."
10. Kohlberg, *Essays on Moral Development*.
11. Aaron, *In Their Own Words*, 37–70.
12. Department of Defense, *Measuring Stability and Security in Iraq*, 2008.
13. Rieger, "Stability and Different Types of Radicalism," 88–91.
14. Thomas and Kiser, "Lords of the Silk Route."

Bibliography

Aaron, David. *In Their Own Words: Voices of Jihad*. Santa Monica, CA: Rand Corporation, 2008.

Aristotle. *Nicomachean Ethics*. Translated by W. D. Ross, 1908. Available at the Constitution Society, http://www.constitution.org/ari/ethic_00.htm.

Babcock, Linda, and George Loewenstein. "Explaining the Bargaining Impasse: The Role of Self-Serving Biases." In *Exotic Preferences: Behavioral Economics and Human Motivation*, edited by George Loewenstein, 215–40. New York: Oxford University Press, 2007.

Department of Defense. *Measuring Stability and Security in Iraq*. Report to Congress, in accordance with the Department of Defense Appropriations Act 2008, Sections 9010 and 9204, September 2008.

Kohlberg, Lawrence. *Essays on Moral Development*. Vol. 1, *The Philosophy of Moral Development*. San Francisco: Harper & Row, 1981.

Loewenstein, George. "Behavioral Decision Theory and Business Ethics: Skewed Trade-Offs between Self and Other." In *Codes of Conduct: Behavioral Research into Business Ethics*, edited by David M. Messick and Ann E. Tenbrunsel, 214–27. New York: Russel Sage, 1996.

National Counterterrorism Center. *2007 Report on Terrorism*, 30 April 2008.

Palomino, Maria L., and Marco Aquino. "Peru Police Held Hostage as Mining Unrest Deepens." *El Economista*, 17 June 2008.

Rate, Christopher R. "What Is Courage? A Search for Meaning." PhD diss., Yale University, 2007.

Rieger, Thomas. "Stability and Different Types of Radicalism." In *Anticipating Rare Events: Can Acts of Terror, Use of Weapons of Mass Destruction or Other High Profile Acts Be Anticipated? A Scientific Perspective on Problems, Pitfalls and Prospective Solutions*, edited by Nancy Chesser, 88–91. Topical Strategic Multi-Layer Assessment (SMA) Multi-Agency/Multi-Disciplinary White Papers in Support of Counter-Terrorism and Counter-WMD, November 2008. http://www.hsdl.org/?view&doc=105192&coll=public.

Snyder, C. R., and Shane J. Lopez. *Positive Psychology: The Scientific and Practical Explorations of Human Strengths*. Thousand Oaks, CA: Sage, 2007.

Thomas, Troy S., and Stephen D. Kiser. "Lords of the Silk Route: Violent Non-State Actors in Central Asia." USAF Institute for National Security Studies (INSS) Occasional Paper 43, May 2002.

Walt, Vivienne. "The World's Growing Food-Price Crisis." *Time*, 27 February 2008. http://www.time.com/time/world/article/0,8599,1717572,00.html.

Chapter 11

Communications

Framing Effects and Political Behavior

Toby Bolsen, Georgia State University

Abstract: While a large body of literature has shown that framing communication affects attitudes and preferences, few studies have demonstrated framing effects on observed behavior. In the few studies that do demonstrate a direct impact of framing on behavior (goal-framing studies), the psychological processes are not well understood. Individual behavior is embedded in a social context. Research on framing effects has helped to identify one specific contextual factor (i.e., rhetoric) that shapes individuals' attitudes and preferences. But attitudes and preferences represent one of several factors known to determine behavior. Moreover, attitudes and preferences vary in terms of the degree of importance people attach to them, and these variations have consequences in terms of promoting action. Recent efforts to identify individual and contextual moderators of framing effects represent an important first step in assessing the external validity of experimental findings. These studies also raise doubts about the pervasive view that framing effects threaten the rational foundations of democratic self-governance.

Framing Effects and Political Behavior

Human behavior is affected by numerous individual and environmental factors. One factor that has received considerable attention among scholars is frames used in communication (e.g., by politicians, journalists, interest groups, etc.). *Framing* is a term that has been loosely used to refer to circumstances in which different words, phrases, and presentation styles affect how a person understands a problem or situation.[1] In this paper, I differentiate among various types of framing effects, explore the psychological processes driving the effects, and assess literature linking frames to individual behavior.

Although scholars know a great deal about how framed messages affect individuals' attitudes, they know much less about the conditions under which these communications impact behavior. The vast majority of studies focus on how framed communications affect attitudes and preferences,[2] which do not always correspond to how a person behaves.[3] Although attitudes can be an important determinant of action, other factors are known to shape intentions and actions (e.g., social norms, values, and habits).[4] Recently social psychologists have begun to explore the direct impact of framing on behavior.[5] I review these studies and assess the normative implications in terms of the criteria on which individuals base their actions.

Classifying Framing Effects

The term *framing* has been used by scholars to refer to distinct phenomena whereby exposure to different types of messages (i.e., substantively different or logically equivalent) shapes individuals' attitudes and preferences.[6] The most commonly encountered frames in political contexts are emphasis frames.[7] An emphasis-framing effect occurs when a speaker uses substantively different words, images, or phrases to influence the way a perceiver thinks about a person, issue, or action. For instance, in deciding whether to allow a local hate rally, officials could emphasize the importance of extending freedom of speech to all groups or the threat that this demonstration might pose to public safety.[8] The frame emphasized in the communication may influence citizens' preferences (i.e., whether to support or oppose the demonstration). A large body of literature shows that similar emphasis frames (e.g., persuasive communications) have a direct effect on political attitudes;[9] however, an important area of future work is to look at the effects of this type of rhetoric on explicit behavior.

A second class of framing effects involves the use of different but logically equivalent words, phrases, and presentation styles to alter individuals' preferences. For instance, whether a policy proposal is framed as resulting in "95 percent employment" or "5 percent unemployment" has been shown to affect policy support. Equivalency-framing effects typically occur when a frame casts the same information in a positive or negative light.[10] These effects violate critical assumptions of a rational framework of choice.[11] To understand why, it is critical to understand what is meant by *preference*. A preference is a rank ordering over a set of comparable objects, and many scholars (e.g., economists) assume that preferences must "possess specific properties including transitivity (e.g., if one prefers chocolate to vanilla, and vanilla to strawberry, then he/she must prefer chocolate to strawberry too) and invariance (e.g., a person's preference should not change if asked whether he/she prefers 'vanilla to chocolate' as compared to being asked whether he/she prefers 'chocolate to vanilla.'"[12] As I explain in detail below, equivalency-framing effects clearly demonstrate a systematic failure of preference invariance because logically equivalent information presented in alternative ways causes individuals to alter their choices. Thus, the existence of equivalency-framing effects has led some to question the normative implications of a system of social choice rooted in the aggregation of attitudes rather than preferences.[13] I address these concerns below.

Levin, Schneider, and Gaeth developed a typology of three different forms of equivalency-framing effects: risky-choice framing, attribute framing, and goal framing.[14] While these three forms of equivalency-framing effects differ in terms of what is framed, the underlying psychological mechanisms driving the effect, and how the effects are detected (i.e., the dependent variable), they each show that logically equivalent communications have differential effects on attitudes, preferences, or actions. One type of equivalency-framing effect is risky-choice framing. In risky-choice framing, two options (i.e., policies) that differ in terms of risk are described in different ways.[15]

A classic example involves a choice between two programs to combat the spread of a disease expected to kill 600 people.[16] In this problem, participants are given a choice between adopting two programs. The experiment involves manipulating the way each policy option is described—for example, in terms of the number of lives that will be saved or the number that will die.

143

One group of subjects is informed that if Program A is adopted, 200 people will be saved, and if Program B is adopted, there is a 1/3 probability that 600 people will be saved and a 2/3 probability that no people will be saved. Described in these terms, Program A represents a risk-averse choice because the outcome is certain (in one instance of this experiment, 72 percent chose this option), and Program B represents a risk-seeking choice because the outcome is uncertain (28 percent chose this option). The other half of the subjects are asked about the same programs in terms of the number of people that would die (Program A = "400 die" and Program B = "1/3 probability nobody will die and 2/3 probability that 600 will die"). When framed in terms of the number of people that would die (negative) rather than the number that would be saved (positive), 78 percent of respondents opted for Program B (the risk-seeking alternative) and only 22 percent selected Program A (the risk-averse option) in the aforementioned instance of this experiment. Thus, aggregate preferences toward identical policies shifted by 50 percent based on equivalent descriptions of the programs. Risky-choice framing effects have received a great deal of attention because of the magnitude of the preference reversals, but the implications of these effects on behavior are unclear, given that the dependent variable is a measure of preference between hypothetical policy alternatives.

A second variant of equivalency framing focuses on the framing of a single attribute of an object or event—attribute framing. For instance, people evaluate beef as tasting better and being less greasy when described as "75 percent lean" instead of "25 percent fat."[17] Similarly, automobiles are evaluated more favorably when citizens are informed of the percentage of American workers employed rather than the percentage of non-American workers employed. In both cases, the frame casts an attribute of a single object in a positive or negative light. Attribute frames differ from risky-choice frames in that (1) a single feature of an object is framed rather than independent options and (2) the effect does not depend on the presence of risk.[18] However, similar to risky-choice frames, attribute-framing effects focus exclusively on how equivalent descriptions of an object's attribute influence attitudes. Future research is necessary to determine whether attribute frames have any effect on behavior.

A third form of equivalency frames causally links framed messages to changes in behavior. Goal framing involves using framed messages to increase the performance of a targeted, desired action. The primary difference between goal-framing effects and other types of equivalency-framing effects is that the dependent variable in these studies is actual behavior rather than policy preferences or attitudes. These studies generally find that the impact of a persuasive message on behavior depends on whether the message stresses the positive benefits of taking action or the negative consequences of *not* taking action.[19] For instance, women are more likely to perform breast self-examination (BSE) when informed that "research shows that women who *do not* do BSE have a *decreased chance* of finding a tumor in the early, more treatable stages," compared to a message saying "research shows that women who *do* engage in BSE have an *increased chance* of finding a tumor in the early, more treatable stages of the disease" (emphasis added).[20] Thus, although these statements are logically equivalent, the negatively charged frame had a more powerful effect than the positive frame in promoting BSE.

Levin, Schneider, and Gaeth explain that goal-framing studies are "relatively new to the framing scene and often involve health-related persuasive messages,"[21] but instances of goal-framing effects can be found in studies ranging from consumer choice to social dilemma problems.[22] Recently, social psychologists have utilized field experiments to explore the effects of these messages on health-related behaviors in the workplace. The purpose is to "address a significant real world problem . . . the low rates of adherence to behavioral recommendations that could reduce the incidence or the mortality from certain cancers or the spread of HIV/AIDS."[23] In two separate field experiments conducted by researchers at the Health, Emotion, and Behavior Lab (HEB) at Yale University, researchers explored the effect of goal-framed messages emphasizing either the "benefits of mammography" or the "risks of neglecting mammography."[24] Twelve months after exposure to the messages, women who viewed the message focusing on the risks of not detecting cancer early were significantly more likely to have obtained a mammography than the women who watched a video emphasizing the benefits of early detection. These findings parallel the results from the Meyerowitz and Chaiken study on BSE.[25] In both cases, a message emphasizing the negative effects of not taking action led to greater performance of the behavior.

In several recent studies on health-related goal frames, HEB researchers found that positive goal frames were more likely to promote prevention-related behaviors.[26] In one study, beach-goers were given a brochure either listing the benefits of wearing sunscreen for preventing skin cancer or focusing on the increased risk of cancer from not wearing protective sunscreen.[27] In this case, the positive message was the stronger motivator of action. While the findings here seem to contradict the two goal-framing studies described above—because negative messages increased BSE and mammograms and positive messages increased sunscreen application—this research shows that it is crucial to differentiate between prevention-related (e.g., applying sunscreen) and detection-related (e.g., performing BSE, obtaining a mammogram, etc.) behaviors in terms of the impact of goal frames on this domain of actions.[28] Detection-related health actions inherently contain more risk than prevention-related behaviors, and this difference affects the impact of the framed message on behavior. Clearly more research is necessary to determine the processes driving these effects and whether these effects occur in political contexts.

The Psychology of Framing Effects

The various classes of framing effects identified above are driven by distinct psychological processes. Emphasis framing involves altering the salience and/or perceived strength of accessible considerations about an issue.[29] Most scholars agree that emphasis-framing effects occur regularly in political contexts, and they have been characterized as essential to the opinion-formation process.[30] Moreover, competing emphasis frames are "a defining element of most political contexts," and, in the presence of messages highlighting alternative considerations, individuals consciously evaluate the persuasiveness of opposing sides.[31] Thus, emphasis framing is driven by a conscious evaluation of the perceived strength of accessible considerations.

In contrast, equivalency-framing effects have been treated (erroneously) as a relatively homogenous set of phenomena explained by prospect theory.[32] Prospect theory explains that individuals' preferences are reference dependent—that is, they depend on whether choice options are cast in a positive or negative light (i.e., gains or losses).[33] The theory is based on the observation that individuals tend to be risk averse with gains and risk seeking with losses (as the results of the disease experiment illustrate). While prospect theory describes the pattern of results observed in risky-choice framing problems, it does not explain *why* equivalency-framing effects regularly occur in the absence of risk.[34]

To fill this gap, social and political psychologists have begun exploring the underlying processes driving equivalency-framing effects.[35] Like emphasis framing, equivalency framing alters the accessibility, or salience, of different aspects of information. However, in contrast to emphasis framing, equivalency framing operates subconsciously by increasing positive or negative associations about the object framed.[36] For instance, positive or negative labeling of an object attribute (e.g., percent fat versus percent lean) automatically generates positive or negative associations about the object (e.g., tastiness of beef). Thus, attribute framing is the result of information being encoded based on its descriptive valence and "stimulus-response compatibility effects" in which individuals attend to positive or negative aspects of messages when forming an evaluation.[37]

Unfortunately, the theoretical literature on the mechanisms driving goal-framing effects is underdeveloped. In part, this stems from the fact that goal-framing effects are a relatively recent discovery. Some researchers have tried to explain goal-framing effects in terms of prospect theory; however, this can be problematic when risk levels of alternative actions are not clearly specified.[38] For instance, Meyerowitz and Chaiken explain the results from their BSE study as consistent with prospect theory.[39] However, Levin, Schneider, and Gaeth argue that this relies on the questionable assumption that performing BSE is riskier than not performing BSE.[40] In this problem, and in other similar studies of the effects of health-related messages on behavior, it is often not clear what represents the riskiest course of action. One implication of these studies is that it

147

may be difficult to predict the impact of goal-framing messages on behaviors that do not contain the same level of risk to one's personal health.

Meyerowitz and Chaiken offer a second explanation for the behavioral effect observed in their study—a negativity bias in information processing.[41] It has been well documented that individuals pay greater attention to and are influenced more by negative information than positive information.[42] However, most of this research has focused exclusively on emphasis-framing effects—for example, evaluating the impacts of negative campaign advertisements on voters.[43] Future research must explore the conditions under which these effects occur and whether they can be explained by existing theory.

Linking Framing to Political Behavior

Although social scientists know a great deal about how different types of frames influence attitudes and preferences, much less is known about when and to what degree framing influences behavior. Nearly all the research in political science focuses on the impact of frames on attitudes and reported intentions.[44] However, attitudes have been shown to vary much in terms of their influence on behavior.[45] A key finding is that the strength of the evaluative association toward an object (i.e., an attitude's strength) is the primary factor determining whether an attitude will predict behavior.[46] Persuasive rhetoric (e.g., emphasis frames) has the potential to influence an attitude's strength.[47] However, moving forward it is important to determine whether framing effects influence behavior by altering specific constructs related to attitude strength—for example, by increasing the perceived importance of an action. Most research in political science focuses on the process by which frames heighten the accessibility of considerations toward an object, and an attitude's accessibility has been only weakly linked to political behavior.[48] Other strength-related characteristics of attitudes, such as an attitude's importance and the degree of certainty with which it is held, have been linked more closely to political action—for example, turning out to vote and issue-based voting in elections.[49]

Equivalency framing may be less important to understand given that these effects are difficult to locate in real political contexts.[50] Nonetheless, this class of framing effects has been cited as a clear demonstration of the limitations of human rationality because a necessary assumption of rational-choice theories is that different representations of an identical choice should yield the same preference.[51] Some go so far as to cite equivalency-framing effects as evidence that citizens do not have real preferences because they are influenced by arbitrary features of how a choice is described.[52] If citizens lack preferences, it begs the question of whether people are capable of self-governance.

While equivalency-framing effects may be normatively disconcerting, it seems premature to make generalizations about their implications for the rational underpinnings of political behavior. These effects have been documented in isolated settings and on specific types of problems. One difficulty for researchers interested in studying the effects of equivalency framing on behavior is that they are nearly impossible to find in real political contexts. The presence of competitive elite frames, political discussion networks, and other contextual cues (e.g., party and interest group positions) suggests that equivalency frames may not be encountered too often in the social world. If equivalency-framing effects occur primarily inside of a laboratory but are moderated by internal characteristics and the social context, then they represent little more than "probability puzzles" that demonstrate isolated violations of rationality.[53] However, if these effects are more pervasive across contexts, then proponents of democracy must confront the normative implications of a system of social choice rooted in attitudes as opposed to preferences.[54]

Moderators of Framing Effects

A critical point about psychological processes driving equivalency framing is that "under certain conditions, individuals do not assimilate accessible information (i.e., do not focus on the negative or positive valence of the information)."[55] Individual characteristics and cues from the social context may interrupt the "accessibility assimilation process" and "moderate accessibility processes by leading individuals to resist the impact of the initial frame, envision alternative frames, and, as a result, avoid being driven by a particular frame."[56] This literature shows that framing effects are less likely when a respondent is male, has high cognitive ability, has strong attitudes or personal involvement with an issue, briefly thinks about his or her decision, or is asked to offer a rationale for his or her decision.[57]

Framing studies have also been criticized for failing to account for the influence of the social context on behavior.[58] Several recent studies offer clear evidence of the power of the social context to moderate framing effects. For instance, Druckman replicated the disease problem, but instead of labeling the choice options Program A and Program B, he labeled them the "Democrats' Program" or the "Republicans' Program."[59] In this example, party cues significantly reduced the risky-choice framing effect. Druckman concludes that "the importance of these results is that the political context leads people to base their preferences on systematic information rather than on arbitrary information contained in the frames."[60] In another study, Druckman explored two other potential moderators of equivalency-framing effects that are features of most political contexts: (1) the presence of alternative frames and (2) the presence of deliberation. In the study, individuals responded to four randomly ordered equivalency-framing problems (two involving risk and two not involving risk).[61] He found that counterframes (i.e., presenting both frames) eliminated equivalency-framing effects entirely and deliberation reduced these effects significantly.[62]

More recently, noncognitive factors such as emotions have been shown to affect individuals' susceptibility to risky-choice frames.[63] Interestingly, distinct emotions operate in different ways: enthusiasm increases risk-seeking behavior because individuals are more likely to anticipate positive outcomes (i.e., optimistic appraisal of risk); anger leads to the adoption of a negative frame and greater risk-taking; and distress enhances the susceptibility to both positive and negative frames.[64] Druckman and McDermott also examined the impact of emotions on preference confidence in the context of the disease problem.[65] They found that anger increases confidence in individuals' preferences, while distress tends to reduce preference confidence. This suggests a possible link among mood, framing effects, and behavior. Emotion may be linked to behavior through an "affective intelligence" system that operates in parallel to cognitive processing.[66] In the case of risky-choice framing, action may be influenced by an affect heuristic that short-circuits cognitive processing of information in the presence of risk.[67] Future research on emotions and framing may shed light on the affective processes linking these effects to behavior.

Conclusion

While a large body of literature has shown that framing communication affects attitudes and preferences, few studies have demonstrated framing effects on observed behavior. In the few studies that do demonstrate a direct impact of framing on behavior (goal-framing studies), the psychological processes are not well understood. Individual behavior is embedded in a social context. Research on framing effects has helped to identify one specific contextual factor (e.g., rhetoric) that shapes individuals' attitudes and preferences. But attitudes and preferences represent two of several factors known to determine behavior. Moreover, attitudes and preferences vary in terms of the degree of importance people attach to them, and these variations have consequences in terms of promoting action. Recent efforts to identify individual and contextual moderators of framing effects represent an important first step in assessing the external validity of experimental findings. These studies also raise doubts about the pervasive view that framing effects threaten the rational foundations of democratic self-governance.

Notes

(All notes appear in shortened form. For full details, see the appropriate entry in the bibliography.)

1. Druckman, "Implications of Framing Effects."
2. Attitudes are defined as a summary evaluation toward an object—for example, a policy, candidate, political issue, and so forth (Chong and Druckman, "Framing Theory"; Fazio, "Attitudes as Object-Evaluation Associations"). Preferences refer to "comparative evaluations (i.e., a ranking over) a set of objects" (Druckman and Lupia, "Preference Formation," 2).
3. LaPiere, "Attitudes versus Actions"; Wicker, "Attitudes versus Actions"; and Fazio, "Attitudes as Object-Evaluation Associations."
4. Ajzen and Fishbein, "Influence of Attitudes on Behavior."

5. Rothman et al., "Strategic Use of Gain- and Loss-Framed Messages"; Salovey and Williams-Piehota, "Field Experiments in Social Psychology"; and Levin, Schneider, and Gaeth, "All Frames Are Not Created Equal."

6. Druckman, "What's It All About?"; Druckman, "Implications of Framing Effects"; Kuhberger, "Influence of Framing on Risky Decisions"; and Levin, Schneider, and Gaeth, "All Frames Are Not Created Equal."

7. Druckman, "Implications of Framing Effects," 226–28.

8. Nelson, Oxley, and Clawson, "Toward a Psychology of Framing Effects."

9. O'Keefe, *Persuasion: Theory and Research*; Druckman and Parkin, "Impact of Media Bias"; and Chong and Druckman, "Framing Theory."

10. Druckman, "Implications of Framing Effects," 228. Social psychologists sometimes refer to equivalency framing effects as "valence" or "message" framing. These terms refer to the same phenomenon, in which logically equivalent messages alter preferences and actions.

11. Kahneman, "Preface."

12. Druckman, "What's It All About?" 2.

13. Bartels, "Democracy with Attitudes."

14. Levin, Schneider, and Gaeth, "All Frames Are Not Created Equal."

15. Ibid., 150.

16. Tversky and Kahneman, "Framing of Decisions"; and Tversky and Kahneman, "Prospect Theory."

17. Levin and Gaeth, "Framing of Attribute Information," 159.

18. Levin, Schneider, and Gaeth, "All Frames Are Not Created Equal," 166.

19. Ibid., 168; and Rothman et al., "Strategic Use of Gain- and Loss-Framed Messages."

20. Meyerowitz and Chaiken, "Effect of Message Framing on Breast Self-Examination Attitude," 504.

21. Levin, Schneider, and Gaeth, "All Frames Are Not Created Equal," 168.

22. For example, consumers are more willing to forgo a cash discount (gain frame) than pay a credit card surcharge (loss frame) (Thaler, "Toward a Positive Theory of Consumer Choice"; and Ganzach and Karsahi, "Message Framing and Buying Behavior"). Similarly, in social dilemma problems, individuals are more willing to refrain from using a "common resource" (i.e., forgo a benefit) than to contribute to the provision of a "public good" (i.e., suffer a personal loss) (Brewer and Kramer, "Choice Behavior in Social Dilemmas"; and Fleishman, "Effects of Decision Framing and Others' Behavior").

23. Salovey and Williams-Piehota, "Field Experiments in Social Psychology," 489.

24. Banks et al., "Effects of Message Framing on Mammography Utilization"; and Schneider et al., "Effects of Message Framing and Ethnic Targeting."

25. Meyerowitz and Chaiken, "Effect of Message Framing on Breast Self-Examination Attitudes."

26. Salovey and Williams-Piehota, "Field Experiments in Social Psychology"; and Rothman et al., "Systematic Influence of Gain- and Loss-Framed Messages."

27. Rothman et al., "Systematic Influence of Gain- and Loss-Framed Messages."

28. Ibid.; Rothman et al., "Strategic Use of Gain- and Loss-Framed Messages"; and Apanovitch et al., "Using Message Framing to Motivate HIV Testing."

29. Iyengar and Kinder, *News That Matters*; Zaller, *Nature and Origin of Mass Opinion*; and Druckman, "Implications of Framing Effects."

30. Druckman, "Implications of Framing Effects," 235; and Chong, "How People Think, Reason, and Feel about Rights," 870.

31. Chong and Druckman, "Framing Theory," 101.

32. Levin, Schneider, and Gaeth, "All Frames Are Not Created Equal"; and Kuhberger, "Influence of Framing on Risky Decisions."

33. Tversky and Kahneman, "Prospect Theory"; and Tversky and Kahneman, "Framing of Decisions and the Psychology of Choice."

34. Druckman, "Political Preference Formation," 674.

35. Druckman, "Political Preference Formation"; Levin, Schneider, and Gaeth, "All Frames Are Not Created Equal"; and Jou, Shanteau, and Harris, "Information Processing View of Framing Effects."

36. Druckman, "Political Preference Formation," 674.

37. Levin, Schneider, and Gaeth, "All Frames Are Not Created Equal," 164.

38. Ibid., 176.

39. Meyerowitz and Chaiken, "Effect of Message Framing on Breast Self-Examination Attitudes."

40. BSE is said to be riskier because the negative frame (i.e., "decreased chance of finding a tumor") is increasing "risk-seeking" behavior (i.e., engaging is BSE) (Levin, Schneider, and Gaeth, "All Frames Are Not Created Equal").

41. Meyerowitz and Chaiken, "Effect of Message Framing on Breast Self-Examination Attitudes."

42. Fiske and Taylor, *Social Cognition*; Taylor, "Asymmetrical Effects of Positive and Negative Events"; Ansolabehere and Iyengar, *Going Negative*; and Geer, *In Defense of Negativity*.

43. For a meta review, see Lau, Sigelman, and Rovner, "Effects of Negative Political Campaigns."

44. But see Lau and Redlawsk, "Advantages and Disadvantages of Cognitive Heuristics"; Miller et al., "Impact of Policy Change Threat"; and Brader, Valentino, and Suhay, "What Triggers Public Opposition to Immigration?"

45. For a review, see Fazio, "Attitudes as Object-Evaluation Associations"; and Ajzen and Fishbein, "Influence of Attitudes on Behavior."

46. Krosnick and Petty, *Attitude Strength*; Visser, Krosnick, and Simmons, "Distinguishing the Cognitive and Behavioral Consequences"; and Visser, Bizer, and Krosnick, "Exploring the Latent Structure."

47. O'Keefe, *Persuasion*; and Chong and Druckman, "Framing Theory."

48. Miller and Peterson, "Theoretical and Empirical Implications of Attitude Strength."

49. Visser, Krosnick, and Simmons, "Distinguishing the Cognitive and Behavioral Consequences"; Krosnick, "Role of Attitude Importance in Social Evaluation"; and Schuman and Presser, *Questions and Answers in Attitude Surveys*.

50. Druckman, "Political Preference Formation."

51. Ibid., 671; and Tversky and Kahneman, "Rational Choice," S253. Emphasis framing effects do not violate assumptions of rational decision making because they occur when substantively different considerations (i.e., new information) surface. This new information can shape the relative prominence of alternative considerations.

52. Bartels, "Democracy with Attitudes," 14.

53. Riker, "Political Psychology of Rational Choice Theory," 32.

54. Bartels, "Democracy with Attitudes."

55. Druckman, "Political Preference Formation," 674.

56. Ibid.

57. Cf. Druckman, "Implications of Framing Effects," 236; see also Miller and Fagley, "Effects of Framing"; Fagley and Miller, "Framing Effects and Arenas of Choice"; Levin, Schneider, and Gaeth, "All Frames Are Not Created Equal"; Takemura, "Influence of Elaboration"; and Sieck and Yates, "Exposition Effects on Decision Making."

58. Druckman and Lupia, "Preference Formation"; Stern, "Toward a Coherent Theory"; Druckman, "Political Preference Formation"; and Bolsen, "Light Bulb Goes On."

59. Druckman, "Implications of Framing Effects," 238.

60. Ibid., 239.

61. Druckman, "Political Preference Formation." Participants were assigned to one of eight conditions that varied the valence of the frame (positive or negative), the presence of counterframes (i.e., both the positively and negatively worded alternatives), and the presence of homogeneous or heterogeneous discussion groups (participants were given a chance to discuss the framing problem they received with three other participants).

62. Ibid., 678. This parallels research on emphasis framing which finds that counterframes (e.g., presenting both sides of an issue) and political discussion tend to reduce the impact of frames. However, discussion groups have distinct effects depending on their composition and the characteristics of individuals in the group (e.g., Druckman and Nelson, "Framing and Deliberation"; and Sniderman and Theriault, "Structure of Political Argument").

63. Druckman and McDermott, "Emotion and the Framing of Risky Choice."

64. Ibid., 11–13.

65. Druckman and McDermott, "Emotion and the Framing of Risky Choice."

66. Marcus, Neuman, and MacKuen, *Affective Intelligence and Political Judgment.*

67. Slovic et al., "Affect, Risk, and Decision-Making"; and Loewenstein et al., "Risk as Feelings."

Bibliography

Ajzen, Icek, and Martin Fishbein. "The Influence of Attitudes on Behavior." In *The Handbook of Attitudes*, edited by Dolores Albarracin, Blair T. Johnson, and Mark P. Zanna, 173–222. Mahwah, NJ: Lawrence Erlbaum Associates, 2005.

Ansolabehere, Stephen, and Shanto Iyengar. *Going Negative: How Political Advertisements Shrink and Polarize the Electorate.* New York: Free Press, 1995.

Apanovitch, Anne Marie, Danielle McCarthy, and Peter Salovey. "Using Message Framing to Motivate HIV Testing among Low-Income, Ethnic Minority Women." *Health Psychology* 22 (2003): 60–67.

Banks, Sara M., Peter Salovey, Susan Greener, Alexander J. Rothman, Anne Moyer, John Beauvais, and Elissa Epel. "The Effects of Message Framing on Mammography Utilization." *Health Psychology* 14 (1995): 178–84.

Bartels, Larry M. "Democracy with Attitudes." In *Electoral Democracy*, edited by Michael Bruce MacKuen and George Rabinowitz, 48–82. Ann Arbor, MI: University of Michigan Press, 2003.

Bolsen, Toby. "A Light Bulb Goes On: Attitudes, Values, Social Norms, and Personal Energy Consumption." Unpublished manuscript, 2009.

Brader, Ted, Nicholas A. Valentino, and Elisabeth Suhay. "What Triggers Public Opposition to Immigration? Anxiety, Group Cues, and Immigration Threat." *American Journal of Political Science* 52, no. 4 (2008): 959–78.

Brewer, Marilynn B., and Roderick M. Kramer. "Choice Behavior in Social Dilemmas: Effects of Social Identity, Group Size, and Decision Framing." *Journal of Personality and Social Psychology* 50 (1986): 542–49.

Chong, Dennis. "How People Think, Reason, and Feel about Rights and Liberties." *American Journal of Political Science* 37 (1993): 867–99.

Chong, Dennis, and James N. Druckman. "Framing Theory." *Annual Review of Political Science* 10 (2007): 103–26.

Druckman, James N. "The Implications of Framing Effects for Citizen Competence." *Political Behavior* 23 (2001): 225–56.

————. "Political Preference Formation: Competition, Deliberation, and the (Ir)relevance of Framing Effects." *American Political Science Review* 98, no. 4 (2004): 671–86.

————. "What's It All About?: Framing in Political Science." In *Perspectives on Framing*, edited by Gideon Keren. New York: Psychology Press/Taylor & Francis, 2009.

Druckman, James N., and Arthur Lupia. "Preference Formation." *Annual Review of Political Science* 3 (2000): 1–24.

Druckman, James N., and Kjersten R. Nelson. "Framing and Deliberation." *American Journal of Political Science* 47, no. 3 (2003): 728–44.

Druckman, James N., and Michael Parkin. "The Impact of Media Bias: How Editorial Slant Affects Voters." *Journal of Politics* 67 (2005): 1030–49.

Druckman, James N., and Rose McDermott. "Emotion and the Framing of Risky Choice." *Political Psychology* 30, no. 3 (2008): 297–321.

Fagley, Nancy S., and Paul M. Miller. "Framing Effects and Arenas of Choice." *Organizational Behavior and Human Decision Processes* 71, no. 3 (1997): 355–73.

Fazio, Russell H. "Attitudes as Object-Evaluation Associations of Varying Strength." *Social Cognition* 25, no. 5 (2007): 603–37.

Fiske, Susan T., and Shelley E. Taylor. *Social Cognition*. 2nd ed. New York: McGraw-Hill, 1991.

Fleishman, John A. "The Effects of Decision Framing and Others' Behavior on Cooperation in a Social Dilemma." *Journal of Conflict Resolution* 32 (1988): 162–80.

Ganzach, Yoav, and Nili Karsahi. "Message Framing and Buying Behavior: A Field Experiment." *Journal of Business Research* 32 (1995): 11–17.

Geer, John G. *In Defense of Negativity: Attack Ads and Presidential Campaigns*. Chicago: University of Chicago Press, 2006.

Iyengar, Shanto, and Donald R. Kinder. *News That Matters: Television and American Opinion*. Chicago: University of Chicago Press, 1987.

Jou, Jerwen, James Shanteau, and Richard Jackson Harris. "An Information Processing View of Framing Effects." *Memory and Cognition* 24, no. 1 (1996): 1–15.

Kahn, Kim Fridkin, and Patrick J. Kenney. *The Spectacle of US Senate Campaigns.* Princeton, NJ: Princeton University Press, 1999.

Kahneman, Daniel. "Preface." In *Choice, Values, and Frames,* edited by Daniel Kahneman and Amos Tversky. New York: Cambridge University Press, 2000.

Kahneman, Daniel, and Amos Tversky. *Choice, Values, and Frames.* New York: Cambridge University Press, 2000.

Krosnick, Jon A. "The Role of Attitude Importance in Social Evaluation: A Study of Policy Preferences, Presidential Candidate Evaluations, and Voting Behavior." *Journal of Personality and Social Psychology* 55, no. 2 (1988): 196–210.

Krosnick, Jon A., and Richard E. Petty, eds. *Attitude Strength: Antecedents and Consequences.* Mahwah, NJ: Lawrence Erlbaum Associates, 1995.

Kuhberger, Anton. "The Influence of Framing on Risky Decisions." *Organizational Behavior and Human Decision Processes* 75 (July 1998): 204–31.

LaPiere, R. T. "Attitudes versus Actions." *Social Forces* 13 (1934): 230–37.

Lau, Richard R., and David P. Redlawsk. "Advantages and Disadvantages of Cognitive Heuristics in Political Decision Making." *American Journal of Political Science* 45, no. 3 (2001): 951–71.

Lau, Richard R., Lee Sigelman, and Ivy Brown Rovner. "The Effects of Negative Political Campaigns: A Meta-Analytic Reanalysis." *Journal of Politics* 69, no. 4 (2007): 1176–1209.

Levin, Irwin P., and Gary J. Gaeth. "Framing of Attribute Information before and after Consuming the Product." *Journal of Experimental Social Psychology* 24 (1988): 520–29.

Levin, Irwin P., Sandra L. Schneider, and Gary J. Gaeth. "All Frames Are Not Created Equal: A Typology and Critical Analysis of Framing Effects." *Organizational Behavior and Human Decision Processes* 76, no. 2 (1998): 149–88.

Levy, Jack S. "Applications of Prospect Theory to Political Science." *Synthese* 135 (May 2003): 215–41.

Loewenstein, George F., Elke U. Weber, Christopher K. Hsee, and Ned Welch. "Risk as Feelings." *Psychological Bulletin* 127, no. 2 (2001): 267–86.

157

Marcus, George E., Russell Neuman, and Michael MacKuen. *Affective Intelligence and Political Judgment.* Chicago: University of Chicago Press, 2000.

Meyerowitz, Beth E., and Shelly Chaiken. "The Effect of Message Framing on Breast Self-Examination Attitudes, Intentions, and Behavior." *Journal of Personality and Social Psychology* 52 (1987): 500–10.

Miller, Joanne M., and David A. M. Peterson. "Theoretical and Empirical Implications of Attitude Strength." *Journal of Politics* 66, no. 3 (2004): 847–67.

Miller, Joanne M., Jon A. Krosnick, Allyson Holbrook, and Laura Lowe. "The Impact of Policy Change Threat on Financial Contributions to Interest Groups." Unpublished manuscript. University of Minnesota, 2002.

Miller, Paul M., and N. S. Fagley. "The Effects of Framing, Problem Variations and Providing Rationale on Choice." *Personality and Social Psychology Bulletin* 17 (1991): 517–22.

Nelson, Thomas E., Zoe M. Oxley, and Rosalee A. Clawson. "Toward a Psychology of Framing Effects." *Political Behavior* 19 (1997): 221–46.

O'Keefe, Daniel J. *Persuasion: Theory and Research.* London: Sage Publications, 2002.

Quattrone, George A., and Amos Tversky. "Contrasting Rational and Psychological Analyses of Political Choice." *American Political Science Review* 82, no. 3 (1988): 719–36.

Riker, William H. "The Political Psychology of Rational Choice Theory." *Political Psychology* 16 (1995): 23–44.

Rivers, Susan E., Peter Salovey, David A. Pizarro, Judith Pizarro, and Tamera R. Schneider. "Message Framing and Pap Test Utilization among Women Attending a Community Health Clinic." *Journal of Health Psychology* 10, no. 1 (2005): 65–77.

Rothman, Alexander J., and Peter Salovey. "Shaping Perception to Motivate Healthy Behavior: The Role of Message Framing." *Psychological Bulletin* 121 (1997): 3–19.

Rothman, Alexander J., Roger D. Bartels, John Wlaschin, and Peter Salovey. "The Strategic Use of Gain- and Loss-Framed Messages to Promote Healthy Behavior: How Theory Can Inform Practice." *Journal of Communication* 56 (2006): S202–20.

Rothman, Alexander J., Steven C. Martino, Brian T. Bedell, Jerusha B. Detweiler, and Peter Salovey. "The Systematic Influence of Gain- and Loss-Framed Messages on Interest in and Use of Different Types of Health Behavior." *Personality and Social Psychology Bulletin* 25, no. 11 (1999): 1355–69.

Salovey, Peter, and Pamela Williams-Piehota. "Field Experiments in Social Psychology." *American Behavioral Scientist* 47, no. 5 (2004): 488–505.

Scheufele, Dietram A. "Framing as a Theory of Media Effects." *Journal of Communication* 49 (1999): 103–22.

Schneider, Tamera R., Peter Salovey, Anne Marie Apanovitch, Judith Pizarro, Danielle McCarthy, Janet Zullo, and Alexander Rothman. "The Effects of Message Framing and Ethnic Targeting on Mammography Use among Low-Income Women." *Health Psychology* 20 (2001): 256–66.

Schuman, Howard, and Stanley Presser. *Questions and Answers in Attitude Surveys: Experiments on Question Form, Wording, and Context.* New York: Academic Press, 1981.

Sieck, Winston, and Frank J. Yates. "Exposition Effects on Decision Making: Choice and Confidence in Choice." *Organizational Behavior and Human Decision Processes* 70 (1997): 207–19.

Slovic, Paul, Melissa Finucane, Ellen Peters, and Donald G. MacGregor. "Affect, Risk, and Decision-Making." *Health Psychology* 24 (2005): S35–S40.

Sniderman, Paul M., and Sean M. Theriault. "The Structure of Political Argument and the Logic of Issue Framing." In *Studies in Public Opinion*, edited by Willem E. Saris and Paul M. Sniderman, 133–64. Princeton, NJ: Princeton University Press, 2004.

Stern, Paul C. "Toward a Coherent Theory of Environmentally Significant Behavior." *Journal of Social Issues* 56, no. 3 (2000): 407–24.

Takemura, Kazuhisa. "Influence of Elaboration on the Framing of Decisions." *Journal of Psychology* 128 (1994): 33–39.

Taylor, Shelley E. "Asymmetrical Effects of Positive and Negative Events: The Mobilization-Minimization Hypothesis." *Psychology Bulletin* 110, no. 1 (1991): 67–85.

Thaler, Richard. "Toward a Positive Theory of Consumer Choice." *Journal of Economic Behavior and Organization* 1 (1980): 39–60.

Tversky, Amos, and Daniel Kahneman. "The Framing of Decisions and the Psychology of Choice." *Science* 211 (1981): 453–58.

———. "Prospect Theory." *Econometrica* 47, no. 1 (1979): 263–91.

———. "Rational Choice and the Framing of Decisions." *Journal of Business* 59, no. 4 (1986): S251–78.

Visser, Penny S., George Y. Bizer, and Jon A. Krosnick. "Exploring the Latent Structure of Strength-Related Attitude Attributes." *Advances in Experimental Social Psychology* 38 (2006): 1–67.

Visser, Penny S., Jon A. Krosnick, and J. P. Simmons. "Distinguishing the Cognitive and Behavioral Consequences of Attitude Importance and Certainty: A New Approach to Testing the Common Factor Hypothesis." *Journal of Experimental Social Psychology* 39 (2003): 118–41.

Wicker, Allan W. "Attitudes versus Actions: The Relationship of Verbal and Overt Behavioral Responses to Attitude Objects." *Journal of Social Issues* 25, no. 4 (1969): 41–78.

Zaller, John. *The Nature and Origin of Mass Opinion.* New York: Cambridge University Press, 1992.

Chapter 12

Decision Science

A Subjective Decision Modeling Approach for Identifying Intent

Elisa Jayne Bienenstock, PhD
Allison Astorino-Courtois, PhD, NSI, Inc.

Abstract: Capability and intent are often identified as the two indicators necessary for a state or nonstate actor to pose a threat to US interests. It is broadly recognized that determining an adversary's capability is a complex endeavor that requires information about the availability of facilities, resources, and the expertise necessary to carry out a nefarious activity. Nevertheless, it is an endeavor for which both the intelligence community and the Department of Defense have tested analytic processes and procedures. The process for determining intent, however, is much less well defined, with "proof" of an actor's intent often resting on analysts' own suppositions and perhaps biases. We argue that this need not be so. Even if not directly measurable at the level of brain activity, intent to do harm should manifest in a number of ways. Providing supporting evidence—if not incontrovertible proof—of intent can and should be pursued as rigorously as that of capability. In this chapter we suggest a subjective decision analysis approach for developing and validating models to analyze and then design hypotheses of an actor's intent. Specifically, the subjective decision analysis approach can serve as an analytic framework for developing inferences regarding which adversaries are incentivized to perform certain actions based on their own perceptions, worldviews, and combined preferences. The same model also might be used to help planners and analysts design "decision probes"—ways to test the actual decision maker's preferences and priorities to evaluate (1) the accuracy of the analytic decision model produced and (2) the weights of the interests he or she considers and which intent those weights convey.

Subjective Decision Analysis

A subjective decision approach involves reconstructing a decision unit's (either an individual or group) choice problem. It is subjective in that the elements of the calculus (e.g., options, interests) represent the perception of reality held by the decision maker modeled, rather than an objective reality provided by the analyst. The approach assumes that within their own reality, decision makers are gain maximizers. That is, they are motivated to satisfy or avoid costs on multiple interests of importance to them. Inputs into the subjective decision model include the decision maker's perceived courses of action and interests/motivations. The data that informs the model can come from subject-matter expert (SME) inputs, intelligence reporting, or even stated policy. It is important to note that the intention of the subjective decision analysis is not to make point predictions on what a decision maker will choose to do but to depict the range of likely decision outcomes associated with various conditions (e.g., personality and context). When used to provide insight regarding intent, the analysis goes one step further by making explicit the likelihood of specific outcomes associated with certain initial conditions.

To prepare the reconstructed decision calculus, the analyst assembles the decision maker's interest, perceived strategy/behavior options, and other context factors based on intelligence and other quantitative, qualitative, and SME input. Decision-maker preference orderings are then derived from construction of search-evaluation (SE) matrixes. The SE matrix is a graphic representation of a multidimensional decision process containing the list of decision options perceived by the decision maker (down column one) judged across the interests considered in making that decision.[1] Once the SE matrix is compiled, the decision options are compared along each interest (i.e., down each interest column) and ordinally ranked according to the degree to which that option is perceived by the decision maker to satisfy each interest. That is, perceived options are ranked not in terms of a numeric value but relative to the other options considered. These single-interest ranks then can be aggregated across the set of interests for each outcome to produce a multidimensional preference ordering for the entire choice set. The

SE matrix thus yields a multi-interest-based assessment of the decision maker's incentives to pursue one behavior option versus another.

A simple example illustrates the technique. Imagine that the autocratic leader of a state is considering whether or not to proliferate special nuclear material in contravention of international norms and the state's own treaty obligations. The subjective decision matrix (table 12.1) represents the choice to proliferate, not proliferate, or postpone the decision given the leader's four main interests: (1) enhancing his country's prestige with international actors opposed to the West, (2) maintaining national security against regional threats, (3) maintaining his own personal wealth, and (4) maintaining sufficiently good relations with his state's major ally.

Table 12.1. SE matrix—incentives to proliferate

	International prestige	National security	Personal wealth	Relations with ally	Overall rank
Proliferate	1*	3	1	3	2 (combined ranks = 8)
Do not proliferate	3	2	3	1	3 (combined ranks = 9)
Do not decide now	2	1	2	2	1 (combined ranks = 7)

* Rankings: 1 = best, 3 = worst. Ties are allowed.

As shown in the matrix, if all of the decision maker's interests are considered equally (i.e., if he is an aggregate utility maximizer), the structure of his motivations (interests) indicates that his optimum option is to postpone his decision. His second best option, however, is to proliferate.

We can gain a number of additional insights from this reconstruction of the decision maker's choice problem. First, we see that there are no dominant choices here (i.e., no single option is better than another across all the interests). Second, we see that the two main motivators of a choice to proliferate are the decision maker's interests in enhanced international prestige and personal wealth. If either of these interests is removed, the

163

value of *proliferate* and *do not proliferate* becomes the same. Only if both of these motivations are eliminated does *non-proliferation* become an attractive option. Applying this type of backward inductive reasoning to collections of previously made decisions with known outcomes provides a means of refining model inputs and assumptions.

Backward Inductive Decision Analysis

While not designed for this purpose, backward inductive analysis of prior decisions provides a means of validating initial conditions and conclusions. Rather than using the decision model to predict a future decision, this approach uses the decision model to evaluate and modulate the inputs to past decisions as a way of evaluating the accuracy of and refining assumptions about options and interests. The backward induction allows the analyst, who knows the outcomes of each decision, to experiment with the interests and priorities that were key to each decision. For instance, if the decision analysis described above was retrospective and the known outcome was proliferation, the analyst could surmise that neither wealth nor prestige was of critical importance in that decision. If the same conclusion emerged for multiple prior decisions, the analyst could, with increasing confidence, begin to assess the relative importance of the decision maker's motivations.

Designing Probe Decisions
for Estimating Intent

Gaining an appreciation for the motives likely driving a choice still leaves us with limited insight into the relationship between those motivations and the decision maker's intention to act on them. To assess intent to act, we need to know more about the relative weights of these motives. For example, would the decision maker be willing to accept less satisfaction on his interest in maintaining good relations with his main ally and/or on national security in return for increased personal wealth? Inductive application of the decision model can provide a method for gauging this type of intent. Rather than using the decision

model solely to evaluate the motives driving a potential decision, the analyst or planner can also use the reconstructed
decision calculus to design "decision probes" to test the relative
weights of the interests as well as the leader's willingness to
make the value trade-offs needed to act on those motives. A key
feature of this approach for assessing intent to do harm is that
the probes used to test the implications of the decision model
may be innocuous; they do not need to bear directly on covert
or nefarious intentions. Often decisions from very different
realms are impacted by identical priorities, preferences, and
interests. In fact, the decisions selected for probing need not be
directly related to the threat in question. This is a benefit because information about these innocuous decisions is likely
easier to obtain than information regarding preferences and
intentions relating to decisions a decision maker wishes to con-
ceal. By gaining a deeper understanding of the underlying motivations for this set of decisions, the analysts can then assess
and update assumptions about priorities and intent.

In the current example, an effective probe decision might be
designed to elicit from the decision maker a choice requiring a
trade-off between his interest in personal wealth and good relations with his major ally—for example, whether or not to accept
a bribe of a certain size in exchange for making a public show
of snubbing the ally's ambassador. What does this tell us? Say
the decision probe is carried out and we observe that the decision maker accepted a relatively small bribe and publicly
snubbed the ambassador. While this does not confirm the accuracy of our decision model, it does provide some additional
evidence to suggest the weight of personal wealth versus ally
relations for this decision maker. Namely, he is willing to trade
at least near-term satisfaction of ally relations for a relatively
low gain in wealth. Returning to the proliferate/do not proliferate decision matrix in table 12.1, we may now suspect that
while there are both incentives and disincentives to proliferate,
it is unlikely that the ally-relations motive would be sufficient
to counter the decision maker's interest in pursuing personal
wealth. While still not directly measuring intent, the analytically
derived probe decisions can provide confidence-enhancing evidence that the decision maker's intent (i.e., his evaluation of

the interplay of his interests) may lean more heavily toward proliferation than not.

Conclusion

In general, we can conceive of at least three forms of intent: (1) unwavering intent that we would perhaps call resolve; (2) contingent intent that includes tipping points ("I will move if he comes any closer"); and (3) anticipatory or attribution-based intent in which an actor takes action in the belief that another will take a certain action. The first is relatively static. The latter two involve some estimation of context and events or anticipation of the interests of others. The simple proliferation decision example presented here assumes the first category of intent. However, assessment of intentions to act contingent on actual or anticipated actions may be made and tested in the same way. Not considered are crisis-related effects such as extreme time pressure, fear, stress, and so forth, which may impede a decision maker's ability to follow an intended course of action.

In this paper, we suggest that whether the decision unit is an individual or a leadership group, its interests and incentives can be used to develop a subjective decision model that can help analysts gain insight into and make more informed inferences about its intent to pursue various actions. Conveniently, many of these interests and preferences also influence other decisions. Because this is the case, it is possible to design probe decisions to collect data that will do two things. First, analytically derived probes will give analysts the opportunity to test some of the implications of their models and improve their accuracy and the reliability of conclusions drawn from them. Second, data collected around probe decisions will help analysts determine the perceptions and interests that are the likely drivers of a particular choice and how these are evaluated by the decision maker to generate an intention to act. That said, it is fair to conclude that gauging intent in the way suggested here would require much research and analytic work. However, we hope that—as has been the case with capability—once begun, a tradecraft would develop around the intent estimation process that would streamline the process, detect and mitigate biases, and continue to enhance our abilities to identify intentions to act.

Note

1. For more detailed explanations of SE matrix construction and theoretical foundations, see Allison Astorino-Courtois, "Clarifying Decisions: Assessing the Impact of Decision Structures on Foreign Policy Choices during the 1970 Jordanian Civil War," *International Studies Quarterly* 42 (1998): 733–54; and Zeev Maoz, *National Choices and International Processes* (Cambridge: Cambridge University Press, 1990).

Abbreviations

BOLD	blood-oxygen-level dependent
BSE	breast self-examination
CFL	compact fluorescent light
COCOM	combatant command
COIN	counterinsurgency
DM	decision maker
dmPFC	dorsomedial prefrontal cortex
DOD	Department of Defense
DO JOC	*Deterrence Operations Joint Operating Concept*
EEG	electroencephalograph
fMRI	functional magnetic resonance imaging
HEB Lab	Health, Emotion, and Behavior Lab
IED	improvised explosive device
INF	Intermediate-Range Nuclear Forces
JP	joint publication
mPFC	medial prefrontal cortex
NIE	national intelligence estimate
OODA	observe, orient, decide, act
PET	positron emission tomography
POLRAD	Political Radicals
pSTS	posterior superior temporal sulcus
SCN	social cognitive neuroscience
SE	search-evaluation
SMA	Strategic Multilayer Assessment
SME	subject matter expert
SNIE	special national intelligence estimate
SOCPAC	Special Operations Command, Pacific
SOP	standard operating procedure
TMS	transcranial magnetic stimulation
ToM	theory of mind
TPJ	temporal parietal junction
TTP	tactics, techniques, and procedures
USSTRATCOM	US Strategic Command

From the Mind to the Feet

Assessing the Perception-to-Intent-to-Action Dynamic

Air University Press Team

Chief Editor
Demorah Hayes

Copy Editor
Sandi Davis

*Cover Art, Book Design,
and Illustrations*
Daniel Armstrong

*Composition and
Prepress Production*
Vivian D. O'Neal

Quality Review
Marvin Bassett

Print Preparation and Distribution
Diane Clark

GPO U.S. GOVERNMENT PRINTING OFFICE : 2011— 724-303